GII00360005

PROCEDURES FOR

# PORT STATE CONTROL 2017

## 2018 EDITION

London, 2018

*Published in 2018 by the*
INTERNATIONAL MARITIME ORGANIZATION
4 Albert Embankment, London SE1 7SR
www.imo.org


Printed and bound by CPI Group (UK) Ltd,
Croydon, CR0 4YY


ISBN 978-92-801-1698-4


| IMO PUBLICATION |
| --- |
| Sales number IC650E |

This publication has been prepared from official documents of IMO, and every effort
has been made to eliminate errors and reproduce the original text(s) faithfully. Readers
should be aware that, in case of inconsistency, the official IMO text will prevail.

# Contents

## Chapter 1 – General

## Chapter 2 – Port State inspections

## Chapter 3 – Contravention and detention

## Chapter 4 – Reporting requirements

# Chapter 5 – Review procedures

# Appendices

# Foreword

The International Maritime Organization (IMO) has always acknowledged that effective implementation and enforcement of the global maritime standards contained in its conventions are primarily the responsibility of flag States.

Nevertheless, the Organization has simultaneously recognized that the exercise of the right to carry out port State control (PSC), as provided for in relevant international conventions, also makes an important contribution to ensuring that global maritime standards are being implemented consistently on ships of varying nationalities. PSC involves the inspection of foreign ships in national port areas to verify that the condition and operation of a ship and its equipment comply with the requirements of international regulations.

The IMO Sub-Committee on Implementation of IMO Instruments (III), recognizing the need for a single comprehensive document to facilitate the work of maritime Administrations in general and PSC inspectors in particular, reviewed and amalgamated existing resolutions and documents on PSC.

This resulted in the adoption of resolution A.1119(30) by the IMO Assembly, on 30 November 2017, which contains, as an annex, the Procedures for port State control, 2017, and revokes resolutions A.1052(27). The Assembly requested the Maritime Safety Committee and the Marine Environment Protection Committee to keep the Procedures under review and to amend them as necessary.

The resolution provides basic guidance to port State control officers (PSCOs) on the conduct of PSC inspections, in order to promote consistency in the conduct of inspections worldwide, and to harmonize the criteria for deciding on deficiencies of a ship, its equipment and the crew, as well as the application of control procedures. In particular, the resolution A.1119(30) contains the Guidelines for port State control officers on the ISM Code (appendix 8), which provide guidance to PSCOs for the harmonized application of related technical or operational deficiencies found in relation to the ISM Code during a PSC inspection and the Guidelines for PSCOs on certification of seafarers, manning and hours of rest (appendix 11), which are intended to provide guidance for a harmonized approach of PSC inspections in compliance with SOLAS regulations V/14 (manning) and I/2 (seafarer certification) and chapter VIII (hours of rest) of the STCW Convention, as amended.

The Organization cooperates with PSC regimes within the framework of resolution A.682(17) on Regional co-operation in the control of ships and discharges. Furthermore, the United Nations General Assembly invited IMO to strengthen its functions with regard to PSC in relation to safety and pollution standards as well as maritime security regulations and, in collaboration with the International Labour Organization, labour standards, so as to promote the implementation of globally agreed minimum standards by all States (resolution 58/240).

# Procedures for port
# State control, 2017

**Resolution A.1119(30)**

*adopted on 6 December 2017*

Procedures for port State control, 2017

THE ASSEMBLY,

RECALLING Article 15(j) of the Convention on the International Maritime Organization regarding the functions of the Assembly in relation to regulations and guidelines concerning maritime safety and the prevention and control of marine pollution from ships,

RECALLING ALSO resolution A.1052(27), by which it adopted the Procedures for port State control, 2011,

RECOGNIZING that efforts by port States have greatly contributed to enhanced maritime safety and security, and prevention of marine pollution,

RECOGNIZING ALSO the need to update the Procedures to take account of the amendments to IMO instruments which have entered into force or have become effective since the adoption of resolution A.1052(27),

HAVING CONSIDERED the recommendations made by the Marine Environment Protection Committee, at its seventieth session, and the Maritime Safety Committee, at its ninety-seventh and ninety-eighth sessions,

1        ADOPTS the Procedures for port State control, 2017, as set out in the annex to the present resolution;

2        INVITES Governments, when exercising port State control, to implement the aforementioned Procedures;

3        REQUESTS the Maritime Safety Committee and the Marine Environment Protection Committee to keep the Procedures under review and to amend them as necessary;

4        REVOKES resolution A.1052(27).

# Chapter 1

## *General*

### 1.1    Purpose

This document is intended to provide basic guidance on the conduct of port State control inspections in support of the control provisions of relevant conventions and parts of the IMO Instruments Implementation Code (III Code) (resolution A.1070(28)) and afford consistency in the conduct of these inspections, the recognition of deficiencies of a ship, its equipment, or its crew, and the application of control procedures.

### 1.2    Application

**1.2.1**    These Procedures apply to ships falling under the provisions of:

    **.1**    the International Convention for the Safety of Life at Sea, 1974, as amended (SOLAS 1974);

    **.2**    the Protocol of 1988 relating to the International Convention for the Safety of Life at Sea, 1974 (SOLAS PROT 1988);

    **.3**    the International Convention on Load Lines, 1966 (LL1966);

    **.4**    the Protocol of 1988 relating to the International Convention on Load Lines, 1966 (LL PROT 1988);

    **.5**    the International Convention for the Prevention of Pollution from Ships, 1973, as modified by the 1978 and 1997 Protocols, as amended (MARPOL);

    **.6**    the International Convention on Standards of Training, Certification and Watchkeeping for Seafarers, 1978, as amended (STCW 1978);

    **.7**    the International Convention on Tonnage Measurement of Ships, 1969 (TONNAGE 1969);

    **.8**    the International Convention on the Control of Harmful Anti-fouling Systems on Ships, 2001 (AFS 2001);

    **.9**    the Convention on the International Regulations for Preventing Collisions at Sea, *1972* (COLREG 1972);

    **.10**    the International Convention on Civil Liability for Oil Pollution Damage, 1969 (CLC 1969);

    **.11**    the Protocol of 1992 to amend the International Convention on Civil Liability for Oil Pollution Damage, 1969 (CLC PROT 1992);

    **.12**    the International Convention on Civil Liability for Bunker Oil Pollution Damage, 2001 (BUNKERS 2001);

    **.13**    the International Convention for the Control and Management of Ships' Ballast Water and Sediments, 2004 (BWM 2004); and

    **.14**    the Nairobi International Convention on the Removal of Wrecks, 2007 (NAIROBI WRC 2007),

    hereafter referred to as the relevant conventions.

**1.2.2**    Ships of non-Parties should be given no more favourable treatment (see section 1.5).

**1.2.3**    For ships below convention size, Parties should apply the procedures in section 1.6.

**1.2.4**    When exercising port State control, Parties should only apply those provisions of the conventions which are in force and which they have accepted.

**1.2.5**    Notwithstanding paragraph 1.2.4, in relation to voluntary early implementation of amendments to the 1974 SOLAS Convention and related mandatory instruments, Parties should take into account the Guidelines on the voluntary early implementation of amendments to the 1974 SOLAS Convention and related mandatory instruments (MSC.1/Circ.1565).

**1.2.6**    If a port State exercises control based on:

    **.1**    International Labour Organization (ILO) Maritime Labour Convention, 2006, as amended (MLC, 2006), guidance on the conduct of such inspections is given in the ILO publication, Guidelines for port State control officers carrying out inspections under the Maritime Labour Convention, 2006; or

    **.2**    ILO Convention No. 147, Merchant Shipping (Minimum Standards) Convention, 1976, or the Protocol of 1996 to the Merchant Shipping (Minimum Standards) Convention, 1976, guidance on the conduct of such inspections is given in the ILO publication, Inspection of labour conditions on board ship: Guide-lines for procedure.

## 1.3    Introduction

**1.3.1**    Under the provisions of the relevant conventions set out in section 1.2 above, the Administration (i.e. the Government of the flag State) is responsible for promulgating laws and regulations and for taking all other steps which may be necessary to give the relevant conventions full and complete effect so as to ensure that, from the point of view of safety of life and pollution prevention, a ship is fit for the service for which it is intended and seafarers are qualified and fit for their duties.

**1.3.2**    In some cases it may be difficult for the Administration to exercise full and continuous control over some ships entitled to fly the flag of its State, for instance those ships which do not regularly call at a port of the flag State. The problem can be, and has been, partly overcome by appointing inspectors at foreign ports and/or authorizing recognized organizations to act on behalf of the flag State Administration.

**1.3.3**    The following control procedures should be regarded as complementary to national measures taken by Administrations of flag States in their countries and abroad and are intended to provide assistance to flag State Administrations in securing compliance with convention provisions in safeguarding the safety of crew, passengers and ships, and ensuring the prevention of pollution.

## 1.4    Provision for port State control

Regulation 19 of chapter I, regulation 6.2 of chapter IX, regulation 4 of chapter XI-1 and regulation 9 of chapter XI-2 of SOLAS 1974, as modified by SOLAS PROT 1988; article 21 of LL 1966, as modified by LL PROT 1988; articles 5 and 6, regulation 11 of Annex I, regulation 16.9 of Annex II, regulation 9 of Annex III, regulation 14 of Annex IV, regulation 9 of Annex V and regulation 10 of Annex VI of MARPOL; article X of STCW 1978; article 12 of TONNAGE 1969, article 11 of AFS 2001 and article 9 of BWM 2004 provide for control procedures to be followed by a Party to a relevant convention with regard to foreign ships visiting their ports. The authorities of port States should make effective use of these provisions for the purposes of identifying deficiencies, if any, in such ships which may render them substandard (see section 3.1) and ensuring that remedial measures are taken.

## 1.5    Ships of non-Parties

**1.5.1**    Article I(3) of SOLAS PROT 1988, article 5(4) of MARPOL, article X(5) of STCW 1978, article 3(3) of AFS 2001 and article 3(3) of BWM 2004 provide that no more favourable treatment is to be given to the ships of countries which are not Party to the relevant Convention. All Parties should, as a matter of principle, apply these Procedures to ships of non-Parties in order to ensure that equivalent surveys and inspections are conducted and an equivalent level of safety and protection of the marine environment is ensured.

**1.5.2**    As ships of non-Parties are not provided with SOLAS, Load Lines, MARPOL, AFS or BWM certificates, as applicable, or the crew members may not hold STCW certificates, the port State control officer (PSCO), taking into account the principles established in these Procedures, should be satisfied that the ship and crew do not present a danger to those on board or an unreasonable threat of harm to the marine environment. If the ship or crew has some form of certification other than that required by a convention, the PSCO may take the form and content of this documentation into account in the evaluation of that ship. The conditions of and on such a ship and its equipment and the certification of the crew and the flag State's minimum manning standard should be compatible with the aims of the provisions of the conventions; otherwise, the ship should be subject to such restrictions as are necessary to obtain a comparable level of safety and protection of the marine environment.

## 1.6    Ships below convention size

**1.6.1**    In the exercise of their functions, the PSCOs should be guided by any certificates and other documents issued by or on behalf of the flag State Administration. In such cases, the PSCOs should limit the scope of inspection to the verification of compliance with those certificates and documents.

**1.6.2**    To the extent a relevant instrument is not applicable to a ship below convention size, the PSCO's task should be to assess whether the ship is of an acceptable standard in regard to safety and the environment. In making that assessment, the PSCO should take due account of such factors as the length and nature of the intended voyage or service, the size and type of the ship, the equipment provided and the nature of the cargo.

## 1.7    Definitions

**1.7.1**    *Bulk carrier:* While noting the definitions in SOLAS regulations IX/1.6 and XII/1.1 and resolution MSC.277(85), for the purposes of port State control, PSCOs should be guided by the ship's type indicated in the ship's certificates in determining whether a ship is a bulk carrier and recognize that a ship which is not designated as a bulk carrier as the ship type on the ship certificate may carry certain bulk cargo as provided for in the above instruments.

**1.7.2**    *Clear grounds:* Evidence that the ship, its equipment, or its crew does not correspond substantially with the requirements of the relevant conventions or that the master or crew members are not familiar with essential shipboard procedures relating to the safety of ships or the prevention of pollution. Examples of clear grounds are included in section 2.4.

**1.7.3**    *Deficiency:* A condition found not to be in compliance with the requirements of the relevant convention.

**1.7.4**    *Detention:* Intervention action taken by the port State when the condition of the ship or its crew does not correspond substantially with the relevant conventions to ensure that the ship will not sail until it can proceed to sea without presenting a danger to the ship or persons on board, or without presenting an unreasonable threat of harm to the marine environment, whether or not such action will affect the normal schedule of the departure of the ship.

**1.7.5**    *Initial inspection:* A visit on board a ship to check both the validity of the relevant certificates and other documents, and the overall condition of the ship, its equipment and its crew (see also section 2.2).

**1.7.6**    *More detailed inspection:* An inspection conducted when there are clear grounds for believing that the condition of the ship, its equipment or its crew does not correspond substantially to the particulars of the certificates.

**1.7.7**    *Port State control officer (PSCO):* A person duly authorized by the competent authority of a Party to a relevant convention to carry out port State control inspections, and responsible exclusively to that Party.

**1.7.8**    *Recognized organization:* An organization which meets the relevant conditions set forth in the Code for Recognized Organizations (RO Code) (MSC.349(92) and MEPC.237(65)), and has been assessed and authorized by the flag State Administration in accordance with provisions of the RO Code to provide the necessary statutory services and certification to ships entitled to fly its flag.

**1.7.9** *Stoppage of an operation:* Formal prohibition against a ship to continue an operation due to an identified deficiency or deficiencies which, singly or together, render the continuation of such operation hazardous.

**1.7.10** *Substandard ship:* A ship whose hull, machinery, equipment or operational safety is substantially below the standards required by the relevant convention or whose crew is not in conformance with the safe manning document.

**1.7.11** *Valid certificates:* A certificate that has been issued, electronically or on paper, directly by a Party to a relevant convention or on its behalf by a recognized organization and contains accurate and effective dates meets the provisions of the relevant convention and to which the particulars of the ship, its crew and its equipment correspond.

## 1.8    Professional profile of PSCOs

**1.8.1**    Port State control should be carried out only by qualified PSCOs who fulfil the qualifications and training specified in section 1.9.

**1.8.2**    When the required professional expertise cannot be provided by the PSCO, the PSCO may be assisted by any person with the required expertise, as acceptable to the port State.

**1.8.3**    The PSCOs and persons assisting them should be free from any commercial, financial, and other pressures and have no commercial interest in the port of inspection, the ships inspected, ship repair facilities or any support services in the port or elsewhere, nor should the PSCOs be employed by or undertake work on behalf of recognized organizations or classification societies.

**1.8.4**    A PSCO should carry a personal document in the form of an identity card issued by the port State and indicating that the PSCO is authorized to carry out the control.

## 1.9    Qualification and training requirements of PSCOs

**1.9.1**    The PSCO should be an experienced officer qualified as flag State surveyor.

**1.9.2**    The PSCO should be able to communicate in English with the key crew.

**1.9.3**    Training should be provided for PSCOs to give the necessary knowledge of the provisions of the relevant conventions which are relevant to the conduct of port State control, taking into account the latest IMO Model Courses for port State control.

**1.9.4**    In specifying the qualifications and training requirements for PSCOs, the Administration should take into account, as appropriate, which of the internationally agreed instruments are relevant for the control by the port State and the variety of types of ships which may enter its ports.

**1.9.5**    PSCOs carrying out inspections of operational requirements should be qualified as a master or chief engineer and have appropriate seagoing experience, or have qualifications from an institution recognized by the Administration in a maritime-related field and have specialized training to ensure adequate competence and skill, or be a qualified officer of the Administration with an equivalent level of experience and training, for performing inspections of the relevant operational requirements.

**1.9.6**    Periodic seminars for PSCOs should be held in order to update their knowledge with respect to instruments related to port State control.

# Chapter 2

## *Port State inspections*

### 2.1 General

**2.1.1** In accordance with the provisions of the relevant conventions, Parties may conduct inspections by PSCOs of foreign ships in their ports.

**2.1.2** Such inspections may be undertaken on the basis of:

  .1  the initiative of the Party;

  .2  the request of, or on the basis of information regarding a ship provided by, another Party; or

  .3  information regarding a ship provided by a member of the crew, a professional body, an association, a trade union or any other individual with an interest in the safety of the ship, its crew and passengers, or the protection of the marine environment.

**2.1.3** Whereas Parties may entrust surveys and inspections of ships entitled to fly their own flag either to inspectors nominated for this purpose or to recognized organizations, they should be aware that, under the relevant conventions, foreign ships are subject to port State control, including boarding, inspection, remedial action and possible detention, only by officers duly authorized by the port State. This authorization of PSCOs may be a general grant of authority or may be specific on a case-by-case basis.

**2.1.4** All possible efforts should be made to avoid a ship being unduly detained or delayed. If a ship is unduly detained or delayed, it should be entitled to compensation for any loss or damage suffered.

### 2.2 Initial inspections

**2.2.1** In the pursuance of control procedures under the relevant conventions, which, for instance, may arise from information given to a port State regarding a ship, a PSCO may proceed to the ship and, before boarding, gain, from its appearance in the water, an impression of its standard of maintenance from such items as the condition of its paintwork, corrosion or pitting or unrepaired damage.

**2.2.2** At the earliest possible opportunity, the PSCO should ascertain the type of ship, year of build and size of the ship for the purpose of determining which provisions of the conventions are applicable.

**2.2.3** On boarding and introduction to the master or the responsible ship's officer, the PSCO should examine the ship's relevant certificates and documents required by the relevant conventions, as listed in appendix 12. PSCOs should note the following:

  .1  certificates may be in hard copy or electronic form;

  .2  where the ship relies upon electronic certificates:

    .1  the certificates and website used to access them should conform with the Guidelines for the use of electronic certificates (FAL.5/Circ.39/Rev.2 and Corr.1);

    .2  specific verification instructions are to be available on the ship; and

    .3  viewing such certificates on a computer is considered as meeting the requirement that certificates be "on board";

  .3  when examining 1969 International Tonnage Certificates, the PSCO should be guided by appendix 10; and

.4    when examining certificates or documentary evidence of seafarers issued in accordance with STCW 1978, the PSCO should be guided by appendix 11; the list of certificates or documentary evidence required under STCW 1978 is also found in table B-I/2 of the STCW Code.

**2.2.4**    If the certificates required by the relevant conventions are valid and the PSCO's general impression and visual observations on board confirm a good standard of maintenance, the PSCO should generally confine the inspection to reported or observed deficiencies, if any.

**2.2.5**    In conducting an initial inspection, the PSCO should check both the validity of the relevant certificates and other documents (with reference to appendix 12) required by the relevant conventions and the overall condition of the ship, including its equipment, navigational bridge, decks including forecastle, cargo holds/areas, engine-room and pilot transfer arrangements.

**2.2.6**    In pursuance of control procedures under chapter IX of SOLAS 1974 in relation to the International Management Code for the Safe Operation of Ships and for Pollution Prevention (ISM Code), the PSCO should utilize the guidelines in appendix 8.

**2.2.7**    If, however, the PSCO from general impression or observations on board has clear grounds for believing that the ship, its equipment or its crew do not substantially meet the requirements, the PSCO should proceed to a more detailed inspection, taking into consideration sections 2.4 and 2.5. In forming such an impression, the PSCO should utilize the guidelines in relevant appendices.

## 2.3    General procedural guidelines for PSCOs

**2.3.1**    The PSCO should observe the Code of Good Practice for Port State control officers (MSC-MEPC.4/Circ.2), as shown in appendix 1, use professional judgement in carrying out all duties and consider consulting others as deemed appropriate.

**2.3.2**    When boarding a ship, the PSCO should present to the master or to the representative of the owner, if requested to do so, the PSCO identity card. This card should be accepted as documented evidence that the PSCO in question is duly authorized by the Administration to carry out port State control inspections.

**2.3.3**    If the PSCO has clear grounds for carrying out a more detailed inspection, the master should be immediately informed of these grounds and advised that, if so desired, the master may contact the Administration or, as appropriate, the recognized organization responsible for issuing the certificate and invite their presence on board.

**2.3.4**    In the case that an inspection is initiated based on a report or complaint, especially if it is from a crew member, the source of the information should not be disclosed.

**2.3.5**    When exercising control, all possible efforts should be made to avoid a ship being unduly detained or delayed. It should be borne in mind that the main purpose of port State control is to prevent a ship proceeding to sea if it is unsafe or presents an unreasonable threat of harm to the marine environment. The PSCO should exercise professional judgement to determine whether to detain a ship until the deficiencies are corrected or to allow it to sail with certain deficiencies, having regard to the particular circumstances of the intended voyage.

**2.3.6**    It should be recognized that all equipment is subject to failure and spares or replacement parts may not be readily available. In such cases, undue delay should not be caused if, in the opinion of the PSCO, safe alternative arrangements have been made.

**2.3.7**    Where the grounds for detention are the result of accidental damage suffered to a ship, no detention order should be issued, provided that:

.1    due account has been given to the convention requirements regarding notification to the flag State Administration, the nominated surveyor or the recognized organization responsible for issuing the relevant certificate;

.2     prior to entering a port, the master or company has submitted to the port State Authority details of the circumstances of the accident and the damage suffered and information about the required notification of the flag State Administration;

.3     appropriate remedial action, to the satisfaction of the port State Authority, is being taken by the ship; and

.4     the port State Authority has ensured, having been notified of the completion of the remedial action, that deficiencies which were clearly hazardous to safety, health or environment have been rectified.

**2.3.8**   Since detention of a ship is a serious matter involving many issues, it may be in the best interest of the PSCO to act together with other interested parties (see paragraph 4.1.3). For example, the officer may request the owner's representatives to provide proposals for correcting the situation. The PSCO should also consider cooperating with the flag State Administration's representatives or the recognized organization responsible for issuing the relevant certificates, and consulting them regarding their acceptance of the owner's proposals and their possible additional requirements. Without limiting the PSCO's discretion in any way, the involvement of other parties could result in a safer ship, avoid subsequent arguments relating to the circumstances of the detention and prove advantageous in the case of litigation involving "undue delay".

**2.3.9**   Where deficiencies cannot be remedied at the port of inspection, the PSCO may allow the ship to proceed to another port, subject to any appropriate conditions determined. In such circumstances, the PSCO should ensure that the competent authority of the next port of call and the flag State are notified.

**2.3.10**  Detention reports to the flag State should be in sufficient detail for an assessment to be made of the severity of the deficiencies giving rise to the detention.

**2.3.11**  The company or its representative have a right of appeal against a detention taken by the Authority of a port State. The appeal should not cause the detention to be suspended. The PSCO should properly inform the master of the right of appeal.

**2.3.12**  To ensure consistent enforcement of port State control requirements, PSCOs should carry an extract of section 2.3 (General procedural guidelines for PSCOs) for ready reference when carrying out any port State control inspections.

**2.3.13**  PSCOs should also be familiar with the detailed guidelines given in the appendices to these Procedures.

## 2.4   Clear grounds

**2.4.1**   When a PSCO inspects a foreign ship which is required to hold a convention certificate, and which is in a port or an offshore terminal under the jurisdiction of the port State, any such inspection should be limited to verifying that there are on board valid certificates and other relevant documentation and the PSCO forming an impression of the overall condition of the ship, its equipment and its crew, unless there are "clear grounds" for believing that the condition of the ship or its equipment does not correspond substantially with the particulars of the certificates.

**2.4.2**   "Clear grounds" to conduct a more detailed inspection include but are not limited to:

.1     the absence of principal equipment or arrangements required by the relevant conventions;

.2     evidence from a review of the ship's certificates that a certificate or certificates are clearly invalid;

.3     evidence that documentation required by the relevant conventions and listed in appendix 12 is not on board, is incomplete, is not maintained or is falsely maintained;

.4     evidence from the PSCO's general impressions and observations that serious hull or structural deterioration or deficiencies exist that may place at risk the structural, watertight or weathertight integrity of the ship;

.5     evidence from the PSCO's general impressions or observations that serious deficiencies exist in the safety, pollution prevention or navigational equipment;

.6 information or evidence that the master or crew is not familiar with essential shipboard operations relating to the safety of ships or the prevention of pollution, or that such operations have not been carried out;

.7 indications that key crew members may not be able to communicate with each other or with other persons on board;

.8 the emission of false distress alerts not followed by proper cancellation procedures; and

.9 receipt of a report or complaint containing information that a ship appears to be substandard.

## 2.5    More detailed inspections

**2.5.1**    If the ship does not carry valid certificates, or if the PSCO, from general impressions or observations on board, has clear grounds for believing that the condition of the ship or its equipment does not correspond substantially with the particulars of the certificates or that the master or crew is not familiar with essential shipboard procedures, a more detailed inspection as described in this chapter should be carried out, utilizing relevant appendices.

**2.5.2**    It is not envisaged that all of the equipment and procedures outlined in this chapter would be checked during a single port State control inspection, unless the condition of the ship or the familiarity of the master or crew with essential shipboard procedures necessitates such a detailed inspection. In addition, these Procedures are not intended to impose the seafarer certification programme of the port State on a ship entitled to fly the flag of another Party to STCW or to impose control procedures on foreign ships in excess of those imposed on ships of the port State.

# Chapter 3

## Contravention and detention

### 3.1    Identification of a substandard ship

**3.1.1**    In general, a ship is regarded as substandard if the hull, machinery, equipment or operational safety is substantially below the standards required by the relevant conventions or if the crew is not in conformance with the safe manning document, owing to, inter alia:

> .1    the absence of principal equipment or arrangement required by the conventions;
>
> .2    non-compliance of equipment or arrangement with relevant specifications of the conventions;
>
> .3    substantial deterioration of the ship or its equipment, for example, because of poor maintenance;
>
> .4    insufficiency of operational proficiency, or unfamiliarity of essential operational procedures by the crew; and
>
> .5    insufficiency of manning or insufficiency of certification of seafarers.

**3.1.2**    If these evident factors as a whole or individually make the ship unseaworthy and put at risk the ship or the life of persons on board or present an unreasonable threat of harm to the marine environment if it were allowed to proceed to sea, it should be regarded as a substandard ship. The PSCO should also take into account the guidelines in appendix 2.

### 3.2    Submission of information concerning deficiencies

**3.2.1**    Information that a ship appears to be substandard could be submitted to the appropriate authorities of the port State (see section 3.3) by a member of the crew, a professional body, an association, a trade union or any other individual with an interest in the safety of the ship, its crew and passengers, or the protection of the marine environment.

**3.2.2**    This information should be submitted in writing to permit proper documentation of the case and of the alleged deficiencies. If the information is passed verbally, the filing of a written report should be required, identifying, for the purposes of the port State's records, the individual or body providing the information. The attending PSCO may collect this information and submit it as part of the PSCO's report if the originator is unable to do so.

**3.2.3**    Information which may cause an investigation should be submitted as early as possible after the arrival of the ship, giving adequate time to the authorities to act as necessary.

**3.2.4**    Each Party to the relevant convention should determine which authorities should receive information on substandard ships and initiate action. Measures should be taken to ensure that information submitted to the wrong department should be promptly passed on by such department to the appropriate authority for action.

### 3.3    Port State action in response to alleged substandard ships

**3.3.1**    On receipt of information about an alleged substandard ship or alleged pollution risk, the authorities should immediately investigate the matter and take the action required by the circumstances in accordance with the preceding sections.

**3.3.2**   Authorities which receive information about a substandard ship that could give rise to detention should forthwith notify any maritime, consular and/or diplomatic representatives of the flag State in the area of the ship and request them to initiate or cooperate with investigations. Likewise, the recognized organization which has issued the relevant certificates on behalf of the flag State should be notified. These provisions will not, however, relieve the authorities of the port State, being a Party to a relevant convention, of the responsibility for taking appropriate action in accordance with its powers under the relevant conventions.

**3.3.3**   If the port State receiving information is unable to take action because there is insufficient time or no PSCOs can be made available before the ship sails, the information should be passed to the authorities of the country of the next appropriate port of call, to the flag State and also to the recognized organization in that port, where appropriate.

## 3.4   Responsibilities of port State to take remedial action

If a PSCO determines that a ship can be regarded as substandard as specified in section 3.1 and appendix 2, the port State should immediately ensure that corrective action is taken to safeguard the safety of the ship and passengers and/or crew and eliminate any threat of harm to the marine environment before permitting the ship to sail.

## 3.5   Guidance for the detention of ships

Notwithstanding the fact that it is impracticable to define a ship as substandard solely by reference to a list of qualifying defects, guidance for the detention of ships is given in appendix 2.

## 3.6   Suspension of inspection

**3.6.1**   In exceptional circumstances where, as a result of a more detailed inspection, the overall condition of a ship and its equipment, also taking into account the crew conditions, are found to be obviously substandard, the PSCO may suspend an inspection.

**3.6.2**   Prior to suspending an inspection, the PSCO should have recorded detainable deficiencies in the areas set out in appendix 2, as appropriate.

**3.6.3**   The suspension of the inspection may continue until the responsible parties have taken the steps necessary to ensure that the ship complies with the requirements of the relevant instruments.

**3.6.4**   In cases where the ship is detained and an inspection is suspended, the port State Authority should notify the responsible parties without delay. The notification should include information about the detention, and state that the inspection is suspended until that authority has been informed that the ship complies with all relevant requirements.

## 3.7   Procedures for rectification of deficiencies and release

**3.7.1**   The PSCO should endeavour to secure the rectification of all deficiencies detected.

**3.7.2**   In the case of deficiencies which are clearly hazardous to safety or the environment, the PSCO should, except as provided in paragraph 3.7.3, ensure that the hazard is removed before the ship is allowed to proceed to sea. For this purpose, appropriate action should be taken, which may include detention or a formal prohibition of a ship to continue an operation due to established deficiencies which, individually or together, would render the continued operation hazardous.

**3.7.3**   Where deficiencies which caused a detention, as referred to in paragraph 3.7.2, cannot be remedied in the port of inspection, the port State Authority may allow the ship concerned to proceed to the nearest appropriate repair yard available, as chosen by the master and agreed to by that authority, provided that the conditions agreed between the port State Authority and the flag State are complied with. Such conditions will ensure that the ship should not sail until it can proceed without risk to the safety of the passengers or crew, or risk to other ships, or without presenting an unreasonable threat of harm to the marine environment.

Such conditions may include confirmation from the flag State that remedial action has been taken on the ship in question. In such circumstances the port State Authority should notify the authority of the ship's next port of call, the parties mentioned in paragraph 4.1.4 and any other authority as appropriate. Notification to authorities should be made in the form shown in appendix 14. The authority receiving such notification should inform the notifying authority of action taken and may use the form shown in appendix 15.

**3.7.4**    On the condition that all possible efforts have been made to rectify all other deficiencies, except those referred to in paragraphs 3.7.2 and 3.7.3, the ship may be allowed to proceed to a port where any such deficiencies can be rectified.

**3.7.5**    If a ship referred to in paragraph 3.7.3 proceeds to sea without complying with the conditions agreed to by the Authority of the port of inspection that port State Authority should immediately alert the next port, if known, the flag State and all other authorities it considers appropriate.

**3.7.6**    If a ship referred to in paragraph 3.7.3 does not call at the nominated repair port, the port State Authority of the repair port should immediately alert the flag State and detaining port State, which may take appropriate action, and notify any other authority it considers appropriate.

# Chapter 4

## *Reporting requirements*

### 4.1    Port State reporting

**4.1.1**    Port State authorities should ensure that, at the conclusion of an inspection, the master of the ship is provided with a document showing the results of the inspection, details of any action taken by the PSCO, and a list of any corrective action to be initiated by the master and/or company. Such reports should be made in accordance with the format in appendix 13.

**4.1.2**    Where, in the exercise of port State control, a Party denies a foreign ship entry to the ports or offshore terminals under its jurisdiction, whether or not as a result of information about a substandard ship, it should forthwith provide the master and flag State with reasons for the denial of entry.

**4.1.3**    In the case of a detention, at least an initial notification should be made to the flag State Administration as soon as practicable (see paragraph 2.3.8). If such notification is made verbally, it should be subsequently confirmed in writing. As a minimum, the notification should include details of the ship's name, the IMO number, copies of Forms A and B as set out in appendix 13, time of detention and copies of any detention order. Likewise, the recognized organizations which have issued the relevant certificates on behalf of the flag State should be notified, where appropriate. The parties above should also be notified in writing of the release of detention. As a minimum, this information should include the ship's name, the IMO number, the date and time of release and a copy of Form B as set out in appendix 13.

**4.1.4**    If the ship has been allowed to sail with known deficiencies, the authorities of the port State should communicate all the facts to the authorities of the country of the next appropriate port of call, to the flag State, and to the recognized organization, where appropriate.

**4.1.5**    Parties to a relevant convention, when they have exercised control giving rise to detention, should submit to the Organization reports in accordance with SOLAS regulation I/19, article 11 of MARPOL, article 21 of Load Lines, or article X(3) of STCW. Such deficiency reports should be made in accordance with the form given in appendices 13 or 16, as appropriate, or may be submitted electronically by the port State or a regional PSC regime.

**4.1.6**    Copies of such deficiency reports should, in addition to being forwarded to the Organization, be sent by the port State without delay to the authorities of the flag State and, where appropriate, to the recognized organization which had issued the relevant certificate. Deficiencies found which are not related to the relevant conventions, or which involve ships of non-Parties or below convention size, should be submitted to flag States and/or to appropriate organizations but not to IMO.

**4.1.7**    Relevant telephone numbers and addresses of flag States' headquarters to which reports should be sent as outlined above, as well as addresses of flag State offices which provide inspection services should be provided to the Organization.*

### 4.2    Flag State reporting

**4.2.1**    On receiving a report on detention, the flag State and, where appropriate, the recognized organization through the flag State Administration, should, as soon as possible, inform the Organization of remedial action

---

* Such addresses are available in National contact points for safety and pollution prevention and response (MSC-MEPC.6/Circ.15), which may be amended, the IMO Internet Home Page and the GISIS module on contact points (http://gisis.imo.org/Public).

taken in respect of the detention, which may be submitted electronically by the flag State to GISIS or in a format shown in appendix 17.

**4.2.2**   Relevant telephone numbers and addresses of port State control offices, headquarters and those who provide inspection services should be provided to the Organization.

## 4.3   Reporting of allegations under MARPOL

**4.3.1**   A report on alleged deficiencies or on alleged contravention of the discharge provisions relating to the provisions of MARPOL should be forwarded to the flag State as soon as possible, preferably no later than sixty days after the observation of the deficiencies or contravention. Such reports may be made in accordance with the format in appendices 13 or 16, as appropriate. If a contravention of the discharge provisions is suspected, then the information should be supplemented by evidence of violations which, as a minimum, should include the information specified in parts 2 and 3 of appendices 3 and 4 of these Procedures.

**4.3.2**   On receiving a report on alleged deficiencies or alleged contravention of the discharge provisions, the flag State and, where appropriate, the recognized organization through the flag State Administration, should, as soon as possible, inform the Party submitting the report of immediate action taken in respect of the alleged deficiencies or contravention. That Party and the Organization should, upon completion of such action, be informed of the outcome and details, where appropriate, be included in the mandatory annual report to the Organization.

# Chapter 5

## Review procedures

### 5.1 Report of comments

**5.1.1** In the interest of making information regarding deficiencies and remedial measures generally available, a summary of such reports should be made by the Organization in a timely manner in order that the information can be disseminated in accordance with the Organization's procedures to all Parties to the relevant conventions. In the summary of deficiency reports, an indication should be given of flag State action or whether a comment by the flag State concerned is outstanding.

**5.1.2** The appropriate Committee should periodically evaluate the summary of the deficiency reports in order to identify measures that may be necessary to ensure more consistent and effective application of IMO instruments, paying close attention to the difficulties reported by Parties to the relevant conventions, particularly in respect of developing countries in their capacity as port States.

**5.1.3** Recommendations to address such difficulties, when recognized by the appropriate Committee, should, where appropriate, be incorporated into the relevant IMO instrument and any modifications relating to the procedures and obligations should be made in the port State documentation.

# Appendices

# Appendix 1
## Code of good practice for port State control officers conducting inspections within the framework of the regional memoranda of understanding and agreement on port State control (MSC-MEPC.4/Circ.2)

## Introduction

**1**      This Code provides guidelines regarding the standards of integrity, professionalism and transparency that regional port State control (PSC) regimes expect of all port State control officers (PSCOs) who are involved in or associated with port State control inspections.

## Objective

**2**      The objective of this Code is to assist PSCOs in conducting their inspections to the highest professional level. PSCOs are central to achieving the aims of the regional PSC regime. They are the daily contact with the shipping world. They are expected to act within the law, within the rules of their Government and in a fair, open, impartial and consistent manner.

## Fundamental principles of the Code

**3**      The Code of good practice encompasses three fundamental principles against which all actions of PSCOs are judged: integrity, professionalism and transparency. These are defined as follows:

   **.1**      integrity is the state of moral soundness, honesty and freedom from corrupting influences or motives;

   **.2**      professionalism is applying accepted professional standards of conduct and technical knowledge. For PSCOs standards of behaviour are established by the maritime authority and the general consent of the port State members; and

   **.3**      transparency implies openness and accountability.

**4**      The list of the actions and behaviour expected of PSCOs in applying these principles is set out in the annex to this appendix.

**5**      Adhering to professional standards provides greater credibility to PSCOs and places more significance on their findings.

**6**      Nothing in the Code shall absolve the PSCOs from complying with the specific requirements of the PSC instruments and applicable national laws.

# Annex
## *Code of good practice for port State control officers*

### Actions and behaviour of PSCOs

The PSCOs should:

**1**      use their professional judgement in carrying out their duties;

### Respect

**2**      remember that a ship is a home as well as a workplace for the ship's personnel and not unduly disturb their rest or privacy;

**3**      comply with any ship housekeeping rules such as removing dirty shoes or work clothes;

**4**      not be prejudiced by the race, gender, religion or nationality of the crew when making decisions and treat all personnel on board with respect;

**5**      respect the authority of the master or his deputy;

**6**      be polite but professional and firm as required;

**7**      never become threatening, abrasive or dictatorial or use language that may cause offence;

**8**      expect to be treated with courtesy and respect;

### Conduct of inspections

**9**      comply with all health and safety requirements of the ship and their administration, e.g. wearing of personal protective clothing, and not take any action or cause any action to be taken which could compromise the safety of the PSCO or the ship's crew;

**10**      comply with all security requirements of the ship and wait to be escorted around the ship by a responsible person;

**11**      present their identity cards to the master or the representative of the owner at the start of the inspection;

**12**      explain the reason for the inspections; however, where the inspection is triggered by a report or complaint they must not reveal the identity of the person making the complaint;

**13**      apply the procedures of PSC and the convention requirements in a consistent and professional way and interpret them pragmatically when necessary;

**14**      not try to mislead the crew, for example by asking them to do things that are contrary to the relevant conventions;

**15**      request the crew to demonstrate the functioning of equipment and operational activities, such as drills and not make tests themselves;

**16**      seek advice when they are unsure of a requirement or of their findings rather than making an uninformed decision, for example by consulting colleagues, publications, the flag Administration, the recognized organization;

**17**      where it is safe to do so accommodate the operational needs of the port and the ship;

**18**      explain clearly to the master the findings of the inspection and the corrective action required and ensure that the report of inspection is clearly understood;

**19**      issue to the master a legible and comprehensible report of inspection before leaving the ship;

## Disagreements

**20**    deal with any disagreement over the conduct or findings of the inspection calmly and patiently;

**21**    advise the master of the complaints procedure in place if the disagreement cannot be resolved within a reasonable time;

**22**    advise the master of the right of appeal and relevant procedures in the case of detention;

## Integrity

**23**    be independent and not have any commercial interest in their ports and the ships they inspect or companies providing services in their ports. For example, the PSCOs should not be employed from time to time by companies which operate ships in their ports or the PSCOs should not have an interest in the repair companies in their ports;

**24**    be free to make decisions based on the findings of their inspections and not on any commercial considerations of the port;

**25**    always follow the rules of their administrations regarding the acceptance of gifts and favours, e.g. meals on board;

**26**    firmly refuse any attempts of bribery and report any blatant cases to the maritime authority;

**27**    not misuse their authority for benefit, financial or otherwise; and

## Updating knowledge

**28**    update their technical knowledge regularly.

# Appendix 2
*Guidelines for the detention of ships*

## 1     Introduction

**1.1**     When deciding whether the deficiencies found in a ship are sufficiently serious to merit detention, the PSCO should assess whether:

.1     the ship has relevant, valid documentation; and

.2     the ship has the crew required in the minimum safe manning document.

**1.2**     During inspection, the PSCO should further assess whether the ship and/or crew, throughout its forthcoming voyage, is able to:

.1     navigate safely;

.2     safely handle, carry and monitor the condition of the cargo;

.3     operate the engine-room safely;

.4     maintain proper propulsion and steering;

.5     fight fires effectively in any part of the ship if necessary;

.6     abandon ship speedily and safely and effect rescue if necessary;

.7     prevent pollution of the environment;

.8     maintain adequate stability;

.9     maintain adequate watertight integrity;

.10     communicate in distress situations if necessary; and

.11     provide safe and healthy conditions on board.

**1.3**     If the result of any of these assessments is negative, taking into account all deficiencies found, the ship should be strongly considered for detention. A combination of deficiencies of a less serious nature may also warrant the detention of the ship. Ships which are unsafe to proceed to sea should be detained upon the first inspection, irrespective of the time the ship will stay in port.

## 2     General

The lack of valid certificates as required by the relevant conventions may warrant the detention of ships. However, ships flying the flag of States not a Party to a convention or not having implemented another relevant instrument, are not entitled to carry the certificates provided for by the convention or other relevant instrument. Therefore, absence of the required certificates should not by itself constitute a reason to detain these ships; however, in applying the "no more favourable treatment" clause, substantial compliance with the provisions and criteria specified in these Procedures must be required before the ship sails.

## 3     Detainable deficiencies

To assist the PSCO in the use of these Guidelines, there follows a list of deficiencies, grouped under relevant conventions and/or codes, which are considered to be of such a serious nature that they may warrant the detention of the ship involved. This list is not considered exhaustive, but is intended to give examples of relevant items.

## Areas under the SOLAS Convention

**1** Failure of proper operation of propulsion and other essential machinery, as well as electrical installations.

**2** Insufficient cleanliness of engine-room, excess amount of oily-water mixture in bilges, insulation of piping including exhaust pipes in engine-room contaminated by oil, and improper operation of bilge pumping arrangements.

**3** Failure of the proper operation of emergency generator, lighting, batteries and switches.

**4** Failure of proper operation of the main and auxiliary steering gear.

**5** Absence, insufficient capacity or serious deterioration of personal life-saving appliances, survival craft and launching and recovery arrangements (see also MSC.1/Circ.1490/Rev.1).

**6** Absence, non-compliance or substantial deterioration to the extent that it cannot comply with its intended use of fire detection system, fire alarms, fire-fighting equipment, fixed fire-extinguishing installation, ventilation valves, fire dampers and quick-closing devices.

**7** Absence, substantial deterioration or failure of proper operation of the cargo deck area fire protection on tankers.

**8** Absence, non-compliance or serious deterioration of lights, shapes or sound signals.

**9** Absence or failure of the proper operation of the radio equipment for distress and safety communication.

**10** Absence or failure of the proper operation of navigation equipment, taking the relevant provisions of SOLAS regulation V/16.2 into account.

**11** Absence of corrected navigational charts, and/or all other relevant nautical publications necessary for the intended voyage, taking into account that electronic charts may be used as a substitute for the charts.

**12** Absence of non-sparking exhaust ventilation for cargo pump-rooms.

**13** Serious deficiency in the operational requirements listed in appendix 7.

**14** Number, composition or certification of crew not corresponding with safe manning document.

**15** Non-implementation or failure to carry out the enhanced survey programme in accordance with SOLAS regulation XI-1/2 and the International Code on the Enhanced Programme of Inspections during Surveys of Bulk Carriers and Oil Tankers, 2011 (2011 ESP Code), as amended.

**16** Absence or failure of a voyage data recorder (VDR), when its use is compulsory.

## Areas under the IBC Code

**1** Transport of a substance not mentioned in the Certificate of Fitness or missing cargo information.

**2** Missing or damaged high-pressure safety devices.

**3** Electrical installations not intrinsically safe or not corresponding to the Code requirements.

**4** Sources of ignition in hazardous locations.

**5** Contravention of special requirements.

**6** Exceeding of maximum allowable cargo quantity per tank.

**7** Insufficient heat protection for sensitive products.

**8** Pressure alarms for cargo tanks not operable.

**9** Transport of substances to be inhibited without valid inhibitor certificate.

## Areas under the IGC Code

**1**      Transport of a substance not mentioned in the Certificate of Fitness or missing cargo information.

**2**      Missing closing devices for accommodations or service spaces.

**3**      Bulkhead not gastight.

**4**      Defective air locks.

**5**      Missing or defective quick-closing valves.

**6**      Missing or defective safety valves.

**7**      Electrical installations not intrinsically safe or not corresponding to the Code requirements.

**8**      Ventilators in cargo area not operable.

**9**      Pressure alarms for cargo tanks not operable.

**10**      Gas detection plant and/or toxic gas detection plant defective.

**11**      Transport of substances to be inhibited without valid inhibitor certificate.

## Areas under the Load Lines Convention

**1**      Significant areas of damage or corrosion, or pitting of plating and associated stiffening in decks and hull affecting seaworthiness or strength to take local loads, unless properly authorized temporary repairs for a voyage to a port for permanent repairs have been carried out.

**2**      A recognized case of insufficient stability.

**3**      The absence of sufficient and reliable information, in an approved form, which by rapid and simple means enables the master to arrange for the loading and ballasting of the ship in such a way that a safe margin of stability is maintained at all stages and at varying conditions of the voyage, and that the creation of any unacceptable stresses in the ship's structure is avoided.

**4**      Absence, substantial deterioration or defective closing devices, hatch closing arrangements and watertight/weathertight doors.

**5**      Overloading.

**6**      Absence of, or impossibility to read, draught marks and/or Load Line marks.

## Areas under the MARPOL Convention, Annex I

**1**      Absence, serious deterioration or failure of proper operation of the oily-water filtering equipment, the oil discharge monitoring and control system or the 15 ppm alarm arrangements.

**2**      Remaining capacity of slop and/or sludge tank insufficient for the intended voyage.

**3**      Oil Record Book not available.

**4**      Unauthorized discharge bypass fitted.

**5**      Failure to meet the requirements of regulation 20.4 or alternative requirements specified in regulation 20.7.

## Areas under the MARPOL Convention, Annex II

**1**    Absence of P and A Manual.

**2**    Cargo is not categorized.

**3**    No Cargo Record Book available.

**4**    Unauthorized discharge bypass fitted.

## Areas under the MARPOL Convention, Annex IV

**1**    Absence of valid International Sewage Pollution Prevention Certificate.

**2**    Sewage treatment plant not approved and certified by the Administration.

**3**    Ship's personnel not familiar with disposal/discharge requirements of sewage.

## Areas under the MARPOL Convention, Annex V

**1**    Absence of the garbage management plan.

**2**    No garbage record book available.

**3**    Ship's personnel not familiar with disposal/discharge requirements of garbage management plan.

## Areas under the MARPOL Convention, Annex VI

**1**    Absence of valid IAPP Certificate and where relevant EIAPP Certificates and Technical Files.

**2**    A marine diesel engine, with a power output of more than 130 kW, which is installed on board a ship constructed on or after 1 January 2000, or a marine diesel engine having undergone a major conversion on or after 1 January 2000, which does not comply with the $NO_x$ Technical Code 2008, as amended.

**3**    The sulphur content of any fuel oil used on board ships exceeds the following limits:

   **.1**    3.5% m/m on and after 1 January 2012; and

   **.2**    0.5% m/m on and after 1 January 2020.

**4**    The sulphur content of any fuel used on board exceeds 0.1% m/m on and after 1 January 2015 while operating within a $SO_x$ emission control area, and respectively, as per the provisions of regulation 14.

**5**    An incinerator installed on board the ship on or after 1 January 2000 does not comply with requirements contained in appendix IV to the Annex, or the Standard specifications for shipboard incinerators developed by the Organization (resolution MEPC.244(66)).

**6**    Ship's personnel are not familiar with essential procedures regarding the operation of air pollution prevention equipment.

**7**    Absence of valid IEEC (International Energy Efficiency Certificate).

**8**    Absence of Ship Energy Efficiency Management Plan (SEEMP) specific for the ship (this may form part of the ship's Safety Management System (SMS)).

## Areas under the STCW Convention

**1**    Failure of seafarers to hold appropriate certificates to have a valid dispensation or to provide documentary proof that an application for an endorsement has been submitted to the Administration.

**2**    Failure to comply with the applicable safe manning requirements of the Administration.

**3**      Failure of navigational or engineering watch arrangements to conform to the requirements specified for the ship by the Administration.

**4**      Absence in a watch of a person qualified to operate equipment essential to safe navigation, safety radiocommunications or the prevention of marine pollution.

**5**      Inability to provide for the first watch at the commencement of a voyage and for subsequent relieving watches persons who are sufficiently rested and otherwise fit for duty.

**6**      Failure to provide proof of professional proficiency for the duties assigned to seafarers for the safety of the ship and the prevention of pollution.

## Areas which may not warrant a detention, but where, for example, cargo operations have to be suspended

Failure of the proper operation (or maintenance) of inert gas systems, cargo related gear or machinery should be considered sufficient grounds to stop cargo operation.

# Appendix 3
*Guidelines for investigations and inspections carried out under Annex I of MARPOL*

## Part 1
*Inspection of IOPP Certificate, ship and equipment*

## 1    Ships required to carry an IOPP Certificate

**1.1**    On boarding and introduction to the master or responsible ship's officer, the PSCO should examine the IOPP Certificate, including the attached Record of Construction and Equipment, and the Oil Record Book.

**1.2**    The certificate carries the information on the type of ship and the dates of surveys and inspections. As a preliminary check it should be confirmed that the dates of surveys and inspections are still valid. Furthermore, it should be established if the ship carries an oil cargo and whether the carriage of such oil cargo is in conformity with the certificate (see also paragraph 1.11 of the Record of Construction and Equipment for Oil Tankers).

**1.3**    Through examining the Record of Construction and Equipment, the PSCO may establish how the ship is equipped for the prevention of marine pollution.

**1.4**    If the certificate is valid and the general impression and visual observations on board confirm a good standard of maintenance, the PSCO should generally confine the inspection to reported deficiencies, if any.

**1.5**    If, however, the PSCO from general impressions or observations on board has clear grounds for believing that the condition of the ship or its equipment does not correspond substantially with the particulars of the certificate, a more detailed inspection should be initiated.

**1.6**    The inspection of the engine-room should begin with forming a general impression of the state of the engine-room, the presence of traces of oil in the engine-room bilges and the ship's routine for disposing of oil-contaminated water from the engine-room spaces.

**1.7**    Next a closer examination of the ship's equipment as listed in the IOPP Certificate may take place. This examination should also confirm that no unapproved modifications have been made to the ship and its equipment.

**1.8**    Should any doubt arise as to the maintenance or the condition of the ship or its equipment, then further examination and testing may be conducted as considered necessary. In this respect reference is made to annex 3 to the Survey Guidelines under the Harmonized System of Survey and Certification (HSSC), 2017 (resolution A.1120(30)), as may be amended.

**1.9**    The PSCO should bear in mind that a ship may be equipped over and above the requirements of Annex I of MARPOL. If such equipment is malfunctioning the flag State should be informed. This alone however should not cause a ship to be detained unless the discrepancy presents an unreasonable threat of harm to the marine environment.

**1.10**    In cases of oil tankers, the inspection should include the cargo tank and pump-room area of the ship and should begin with forming a general impression of the layout of the tanks, the cargoes carried, and the routine of cargo slops disposal.

## 2 Ships of non-Parties to MARPOL Annex I and other ships not required to carry an IOPP Certificate

**2.1**    As this category of ships is not provided with an IOPP Certificate, the PSCO should be satisfied with regard to the construction and equipment standards relevant to the ship on the basis of the requirements set out in Annex I of MARPOL.

**2.2**    In all other respects the PSCO should be guided by the procedures for ships referred to in section 1 above.

**2.3**    If the ship has some form of certification other than the IOPP Certificate, the PSCO may take the form and content of this documentation into account in the evaluation of that ship.

## 3 Control

In exercising the control functions, the PSCO should use professional judgement to determine whether to detain the ship until any noted deficiencies are corrected or to allow it to sail with certain deficiencies which do not pose an unreasonable threat of harm to the marine environment. In doing this, the PSCO should be guided by the principle that the requirements contained in Annex I of MARPOL, in respect of construction and equipment and the operation of ships, are essential for the protection of the marine environment and that departure from these requirements could constitute an unreasonable threat of harm to the marine environment.

## Part 2
*Contravention of discharge provisions*

**1**    Experience has shown that information furnished to the flag State as envisaged in appendix 5 of these Procedures is often inadequate to enable the flag State to cause proceedings to be brought in respect of the alleged violation of the discharge requirements. This appendix is intended to identify information which is often needed by a flag State for the prosecution of such possible violations.

**2**    It is recommended that, in preparing a port State report on deficiencies, where contravention of the discharge requirements is involved, the authorities of the coastal or port State be guided by the itemized list of possible evidence as shown in part 3 of this appendix. It should be borne in mind in this connection that:

.1    the report aims to provide the optimal collation of obtainable data; however, even if all the information cannot be provided, as much information as possible should be submitted;

.2    it is important for all the information included in the report to be supported by facts which, when considered as a whole, would lead the port or coastal State to believe a contravention had occurred.

**3**    In addition to the port State report on deficiencies, a report should be completed by a port or coastal State, on the basis of the itemized list of possible evidence. It is important that these reports are supplemented by documents such as:

.1    a statement by the observer of the pollution; in addition to the information required under section 1 of part 3 of this appendix the statement should include considerations which lead the observer to conclude that none of any other possible pollution sources is in fact the source;

.2    statements concerning the sampling procedures both of the slick and on board; these should include location of and time when samples were taken, identity of person(s) taking the samples and receipts identifying the persons having custody and receiving transfer of the samples;

.3    reports of analyses of samples taken of the slick and on board; the reports should include the results of the analyses, a description of the method employed, reference to or copies of scientific documentation attesting to the accuracy and validity of the method employed and names of persons performing the analyses and their experience;

.4   a statement by the PSCO on board together with the PSCO's rank and organization;

.5   statements by persons being questioned;

.6   statements by witnesses; all observations, photographs and documentation should be supported by a signed verification of their authenticity; all certifications, authentications or verifications shall be executed in accordance with the laws of the State which prepares them; all statements should be signed and dated by the person making the statement and, if possible, by a witness to the signing; the names of the persons signing statements should be printed in legible script above or below the signature;

.7   photographs of the oil slick; and

.8   copies of relevant recordings, etc., pages of Oil Record Books, logbooks, discharge.

**4**      The report referred to in paragraphs 2 and 3 should be sent to the flag State. If the coastal State observing the pollution and the port State carrying out the investigation on board are not the same, the State carrying out the latter investigation should also send a copy of its findings to the State observing the pollution and requesting the investigation.

## Part 3
*Itemized list of possible evidence on alleged contravention of the MARPOL Annex I discharge provisions*

## 1      Action on sighting oil pollution

### 1.1      Particulars of ship or ships suspected of contravention

.1   Name of ship

.2   Reasons for suspecting the ship

.3   Date and time (UTC) of observation or identification

.4   Position of ship

.5   Flag and port of registry

.6   Type (e.g. tanker, cargo ship, passenger ship, fishing vessel), size (estimated tonnage) and other descriptive data (e.g. superstructure colour and funnel mark)

.7   Draught condition (loaded or in ballast)

.8   Approximate course and speed

.9   Position of slick in relation to ship (e.g. astern, port, starboard)

.10   Part of the ship from which side discharge was seen emanating

.11   Whether discharge ceased when ship was observed or contacted by radio

### 1.2      Particulars of slick

.1   Date and time (UTC) of observation if different from paragraph 1.1.3

.2   Position of oil slick in longitude and latitude if different from paragraph 1.1.4

.3   Approximate distance in nautical miles from the nearest land

.4   Approximate overall dimension of oil slick (length, width and percentage thereof covered by oil)

.5   Physical description of oil slick (direction and form, e.g. continuous, in patches or in windrows)

.6    Appearance of oil slick (indicate categories)

     –    Category A: Barely visible under most favourable light condition

     –    Category B: Visible as silvery sheen on water surface

     –    Category C: First trace of colour may be observed

     –    Category D: Bright band of colour

     –    Category E: Colours begin to turn dull

     –    Category F: Colours are much darker

.7    Sky conditions (bright sunshine, overcast, etc.), lightfall and visibility (km) at the time of observation

.8    Sea state

.9    Direction and speed of surface wind

.10    Direction and speed of current

## 1.3    Identification of the observer(s)

.1    Name of the observer

.2    Organization with which observer is affiliated (if any)

.3    Observer's status within the organization

.4    Observation made from aircraft/ship/shore/otherwise

.5    Name or identity of ship or aircraft from which the observation was made

.6    Specific location of ship, aircraft, place on shore or otherwise from which observation was made

.7    Activity engaged in by observer when observation was made, for example: patrol, voyage, flight (en route from ... to ...)

## 1.4    Method of observation and documentation

.1    Visual

.2    Conventional photographs

.3    Remote sensing records and/or remote sensing photographs

.4    Samples taken from slick

.5    Any other form of observation (specify)

**Note:** A photograph of the discharge should preferably be in colour. Photographs can provide the following information: that a material on the sea surface is oil; that the quantity of oil discharged does constitute a violation of the Convention; that the oil is being, or has been discharged from a particular ship; and the identity of the ship.

Experience has shown that the aforementioned can be obtained with the following three photographs:

–    details of the slick taken almost vertically down from an altitude of less than 300 m with the sun behind the photographer;

–    an overall view of the ship and "slick" showing oil emanating from a particular ship; and

–    details of the ship for the purposes of identification.

### 1.5 Other information if radio contact can be established

**.1** Master informed of pollution

**.2** Explanation of master

**.3** Ship's last port of call

**.4** Ship's next port of call

**.5** Name of ship's master and owner

**.6** Ship's call sign

## 2 Investigation on board

### 2.1 Inspection of IOPP Certificate

**.1** Name of ship

**.2** Distinctive number or letters

**.3** Port of registry

**.4** Type of ship

**.5** Date and place of issue

**.6** Date and place of endorsement.

**Note:** If the ship is not issued an IOPP Certificate, as much as possible of the requested information should be given.

### 2.2 Inspection of supplement of the IOPP Certificate

**.1** Applicable paragraphs of sections 2, 3, 4, 5 and 6 of the supplement (non-oil tankers)

**.2** Applicable paragraphs of sections 2, 3, 4, 5, 6, 7, 8, 9 and 10 of the supplement (oil tankers)

**Note:** If the ship does not have an IOPP Certificate, a description should be given of the equipment and arrangements on board, designed to prevent marine pollution.

### 2.3 Inspection of Oil Record Book (ORB)

**.1** Copy sufficient pages of the ORB – part I to cover a period of 30 days prior to the reported incident.

**.2** Copy sufficient pages of the ORB – part II (if on board) to cover a full loading/unloading/ballasting and tank cleaning cycle of the ship. Also copy the tank diagram.

### 2.4 Inspection of logbook

**.1** Last port, date of departure, draught forward and aft

**.2** Current port, date of arrival, draught forward and aft

**.3** Ship's position at or near the time the incident was reported

**.4** Spot check if positions mentioned in the logbook agree with positions noted in the ORB

### 2.5 Inspection of other documentation on board

**Other documentation relevant for evidence (if necessary, make copies) such as:**

– recent ullage sheets

– records of monitoring and control equipment.

## 2.6    Inspection of ship

.1    Ship's equipment in accordance with the supplement of the IOPP Certificate

.2    Samples taken. State location on board

.3    Traces of oil in vicinity of overboard discharge outlets

.4    Condition of engine-room and contents of bilges

.5    Condition of oily water separator, filtering equipment and alarm, stopping or monitoring arrangements

.6    Contents of sludge and/or holding tanks

.7    Sources of considerable leakage on oil tankers.

**The following additional evidence may be pertinent:**

.8    Oil on surface of segregated or dedicated clean ballast

.9    Condition of pump-room bilges

.10    Condition of COW system

.11    Condition of IG system

.12    Condition of monitoring and control system

.13    Slop tank contents (estimate quantity of water and of oil).

## 2.7    Statements of persons concerned

If the ORB – part I has not been properly completed, information on the following questions may be pertinent:

.1    Was there a discharge (accidental or intentional) at the time indicated on the incident report?

.2    Is the bilge discharge controlled automatically?

.3    If so, at what time was this system last put into operation and at what time was this system last put on manual mode?

.4    If not, what were date and time of the last bilge discharge?

.5    What was the date of the last disposal of residue and how was disposal effected?

.6    Is it usual to effect discharge of bilge water directly to the sea, or to store bilge water first in a collecting tank? Identify the collecting tank

.7    Have oil fuel tanks recently been used as ballast tanks?

If the ORB – part II has not been properly completed, information on the following questions may be pertinent:

.8    What was the cargo/ballast distribution in the ship on departure from the last port?

.9    What was the cargo/ballast distribution in the ship on arrival in the current port?

.10    When and where was the last loading effected?

.11    When and where was the last unloading effected?

.12    When and where was the last discharge of dirty ballast?

.13    When and where was the last cleaning of cargo tanks?

.14    When and where was the last COW operation and which tanks were washed?

.15    When and where was the last decanting of slop tanks?

**.16** What is the ullage in the slop tanks and the corresponding height of interface?

**.17** Which tanks contained the dirty ballast during the ballast voyage (if ship arrived in ballast)?

**.18** Which tanks contained the clean ballast during the ballast voyage (if ship arrived in ballast)?

**In addition, the following information may be pertinent:**

**.19** Details of the present voyage of the ship (previous ports, next ports, trade)

**.20** Contents of oil fuel and ballast tanks

**.21** Previous and next bunkering, type of oil fuel

**.22** Availability or non-availability of reception facilities for oily wastes during the present voyage

**.23** Internal transfer of oil fuel during the present voyage.

**In the case of oil tankers, the following additional information may be pertinent:**

**.24** The trade the ship is engaged in, such as short/long distance, crude or product or alternating crude/product, lightering service, oil/dry bulk

**.25** Which tanks clean and dirty

**.26** Repairs carried out or envisaged in cargo tanks.

**Miscellaneous information:**

**.27** Comments in respect of condition of ship's equipment

**.28** Comments in respect of pollution report

**.29** Other comments.

## 3 Investigation ashore

### 3.1 Analyses of oil samples

**Indicate method and results of the samples' analyses.**

### 3.2 Further information

Additional information on the ship, obtained from oil terminal staff, tank cleaning contractors or shore reception facilities may be pertinent.

**Note:** Any information under this heading is, if practicable, to be corroborated by documentation such as signed statements, invoices, receipts, etc.

## 4 Information not covered by the foregoing

## 5 Conclusion

**.1** Summing up of the investigator's technical conclusions.

**.2** Indication of applicable provisions of Annex I of MARPOL which the ship is suspected of having contravened.

**.3** Did the results of the investigation warrant the filing of a deficiency report?

## Part 4
*Guidelines for in-port inspection of crude oil washing procedures*

### 1    Preamble

**1.1**    Guidelines for the in-port inspection of crude oil washing (COW) procedures, as called for by resolution 7 of the International Conference on Tanker Safety and Pollution Prevention, 1978, are required to provide a uniform and effective control of crude oil washing to ensure compliance of ships at all times with the provisions of MARPOL.

**1.2**    The design of the crude oil washing installation is subject to the approval of the flag Administration. However, although the operational aspect of crude oil washing is also subject to the approval of the same Administration, it might be necessary for a port State Authority to see to it that continuing compliance with agreed procedures and parameters is ensured.

**1.3**    The COW Operations and Equipment Manual has been so specified that it contains all the necessary information relating to the operation of crude oil washing on a particular tanker. The objectives of the inspection would then be to ensure that the provisions of the Manual dealing with safety procedures and with pollution prevention are being strictly adhered to.

**1.4**    The method of the inspection is at the discretion of the port State Authority and may cover the entire operation or only those parts of the operation which occur when the PSCO is on board.

**1.5**    Inspection will be governed by articles 5 and 6 of MARPOL.

### 2    Inspections

**2.1**    A port State should make the appropriate arrangements so as to ensure compliance with requirements governing the crude oil washing of oil tankers. This is not, however, to be construed as relieving terminal operators and ship owners of their obligations to ensure that the operation is undertaken in accordance with the regulations.

**2.2**    The inspection may cover the entire operation of crude oil washing or only certain aspects of it. It is thus in the interest of all concerned that the ship's records with regard to the COW operations are maintained at all times so that a PSCO may verify those operations undertaken prior to the inspection.

### 3    Ship's personnel

**3.1**    The person in charge and the other nominated persons who have responsibility in respect of the crude oil washing operation should be identified. They must, if required, be able to show that their qualifications meet the requirements, as appropriate, of paragraphs 5.2 and 5.3 of the Revised specifications for the design, operation and control of crude oil washing systems (resolution A.446(XI)), as amended.

**3.2**    The verification may be accomplished by reference to the individual's discharge papers, testimonials issued by the ship's operator or by certificates issued by a training centre approved by an Administration. The numbers of such personnel should be at least as stated in the Manual.

### 4    Documentation

The following documents should be available for inspection:

.1    the IOPP Certificate and the Record of Construction and Equipment, to determine:

.1    whether the ship is fitted with a crude oil washing system as required in regulation 33 of MARPOL Annex I;

.2    whether the crude oil washing system is according to and complying with the requirements of regulations 33 and 35 of MARPOL Annex I;

.3   the validity and date of the Operations and Equipment Manual; and

.4   the validity of the Certificate;

.2   the approved Manual;

.3   the Oil Record Book; and

.4   the Cargo Ship Safety Equipment Certificate to confirm that the inert gas system conforms to regulations contained in chapter II-2 of SOLAS 1974.

## 5   Inert gas system

**5.1**   Inert gas system regulations require that instrumentation shall be fitted for continuously indicating and permanently recording at all times when inert gas is being supplied, the pressure and the oxygen content of the gas in the inert gas supply main. Reference to the permanent recorder would indicate if the system had been operating before and during the cargo discharge in a satisfactory manner.

**5.2**   If conditions specified in the Manual are not being met then the washing must be stopped until satisfactory conditions are restored.

**5.3**   As a further precautionary measure, the oxygen level in each tank to be washed is to be determined at the tank. The meters used should be calibrated and inspected to ensure that they are in good working order. Readings from tanks already washed in port prior to inspection should be available for checking. Spot checks on readings may be instituted.

## 6   Electrostatic generation

It should be confirmed either from the cargo log or by questioning the person in charge that presence of water in the crude oil is being minimized as required by paragraph 6.7 of the Revised Specifications for the design, operation and control of crude oil washing systems (resolution A.446(XI)), as amended.

## 7   Communication

It should be established that effective means of communication exist between the person in charge and the other persons concerned with the COW operation.

## 8   Leakage on deck

PSCOs should ensure that the COW piping system has been operationally tested for leakage before cargo discharge and that the test has been noted in the ship's Oil Record Book.

## 9   Exclusion of oil from engine-room

It should be ascertained that the method of excluding cargo oil from the machinery space is being maintained by inspecting the isolating arrangements of the tank washing heater (if fitted) or of any part of the tank washing system which enters the machinery space.

## 10   Suitability of the crude oil

In judging the suitability of the oil for crude oil washing, the guidance and criteria contained in section 9 of the COW Operations and Equipment Manual should be taken into account.

## 11   Checklist

It should be determined from the ship's records that the pre-crude oil wash operational checklist was carried out and all instruments functioned correctly. Spot checks on certain items may be instituted.

## 12 Wash programmes

**12.1** Where the tanker is engaged in a multiple port discharge, the Oil Record Book would indicate if tanks were crude oil washed at previous discharge ports or at sea. It should be determined that all tanks which will, or may be, used to contain ballast on the forthcoming voyage will be crude oil washed before the ship departs from the port. There is no obligation to wash any tank other than ballast tanks at a discharge port except that each of these other tanks must be washed at least in accordance with paragraph 6.1 of the Revised Specifications for the design, operation and control of crude oil washing systems (resolution A.446(XI)), as amended. The Oil Record Book should be inspected to check that this is being complied with.

**12.2** All crude oil washing must be completed before a ship leaves its final port of discharge.

**12.3** If tanks are not being washed in one of the preferred orders given in the Manual, the PSCO should determine that the reason for this and the proposed order of tank washing are acceptable.

**12.4** For each tank being washed it should be ensured that the operation is in accordance with the Manual in that:

.1 the deck mounted machines and the submerged machines are operating either by reference to indicators, the sound patterns or other approved methods;

.2 the deck mounted machines, where applicable, are programmed as stated;

.3 the duration of the wash is as required; and

.4 the number of tank washing machines being used simultaneously does not exceed that specified.

## 13 Stripping of tanks

**13.1** The minimum trim conditions and the parameters of the stripping operations are to be stated in the Manual.

**13.2** All tanks which have been crude oil washed are to be stripped. The adequacy of the stripping is to be checked by hand dipping at least in the aftermost hand dipping location in each tank or by such other means provided and described in the Manual. It should be ascertained that the adequacy of stripping has been checked or will be checked before the ship leaves its final port of discharge.

## 14 Ballasting

**14.1** Tanks that were crude oil washed at sea will be recorded in the Oil Record Book. These tanks must be left empty between discharge ports for inspection at the next discharge port. Where these tanks are the designated departure ballast tanks they may be required to be ballasted at a very early stage of the discharge. This is for operational reasons and also because they must be ballasted during cargo discharge if hydrocarbon emission is to be contained on the ship. If these tanks are to be inspected when empty, then this must be done shortly after the tanker berths. If a PSCO arrives after the tanks have begun accepting ballast, then the sounding of the tank bottom would not be available. However, an examination of the surface of the ballast water is then possible. The thickness of the oil film should not be greater than that specified in paragraph 4.2.10(b) of the Revised Specifications for the design, operation and control of crude oil washing systems (resolution A.446(XI)), as amended.

**14.2** The tanks that are designated ballast tanks will be listed in the Manual. It is, however, left to the discretion of the master or responsible officer to decide which tanks may be used for ballast on the forthcoming voyage. It should be determined from the Oil Record Book that all such tanks have been washed before the tanker leaves its last discharge port. It should be noted that where a tanker back-loads a cargo of crude oil at an intermediate port into tanks designated for ballast, then it should not be required to wash those tanks at that particular port but at a subsequent port.

**14.3** It should be determined from the Oil Record Book that additional ballast water has not been put into tanks which had not been crude oil washed during previous voyages.

**14.4**     It should be verified that the departure ballast tanks are stripped as completely as possible. Where departure ballast is filled through cargo lines and pumps these must be stripped either into another cargo tank or ashore by the special small diameter line provided for this purpose.

**14.5**     The methods to avoid vapour emission where locally required will be provided in the Manual and they must be adhered to. The PSCO should ensure that this is being complied with.

**14.6**     The typical procedures for ballasting listed in the Manual must be observed. The PSCO should ensure this is being complied with.

**14.7**     When departure ballast is to be shifted, the discharge into the sea must be in compliance with regulations 15 and 34 of Annex I of MARPOL. The Oil Record Book should be inspected to ensure that the ship is complying with this.

# Appendix 4
*Guidelines for investigations and inspections carried out under Annex II of MARPOL*

## Part 1
*Inspection of Certificate (COF or NLS Certificate), ship and equipment*

## 1    Ships required to hold a Certificate

**1.1**    On boarding and after introducing oneself to the master or responsible ship's officer, the PSCO should examine the Certificate of Fitness or NLS Certificate and Cargo Record Book.

**1.2**    The Certificate includes information on the type of ship, the dates of surveys and a list of the products which the ship is certified to carry.

**1.3**    As a preliminary check, the Certificate's validity should be confirmed by verifying that the Certificate is properly completed and signed and that required surveys have been performed. In reviewing the Certificate particular attention should be given to verifying that only those noxious liquid substances which are listed on the Certificate are carried and that these substances are in tanks approved for their carriage.

**1.4**    The Cargo Record Book should be inspected to ensure that the records are up to date. The PSCO should check whether the ship left the previous port(s) with residues of noxious liquid substances on board which could not be discharged into the sea. The book could also have relevant entries from the appropriate authorities in the previous ports. If the examination reveals that the ship was permitted to sail from its last unloading port under certain conditions, the PSCO should ascertain that such conditions have been or will be adhered to. If the PSCO discovers an operational violation in this respect, the flag State should be informed by means of a deficiency report.

**1.5**    If the Certificate is valid and the PSCO's general impressions and visual observations on board confirm a good standard of maintenance, the PSCO should, provided that the Cargo Record Book entries do not show any operational violations, confine the inspection to reported deficiencies, if any.

**1.6**    If, however, the PSCO's general impressions or observations on board show clear grounds for believing that the condition of the ship, its equipment, or its cargo and slops handling operations do not correspond substantially with the particulars of the Certificate, the PSCO should proceed to a more detailed inspection:

> **.1**    initially this requires an examination of the ship's approved Procedures and Arrangements Manual (P and A Manual);
>
> **.2**    the more detailed inspection should include the cargo and pump-room areas of the ship and should begin with forming a general impression of the layout of the tanks, the cargoes carried, pumping and stripping conditions and cargo;
>
> **.3**    next a closer examination of the ship's equipment as shown in the P and A Manual may take place. This examination should also confirm that no unapproved modifications have been made to the ship and its equipment; and
>
> **.4**    should any doubt arise as to the maintenance or the condition of the ship or its equipment then further examination and testing may be conducted as may be necessary. In this respect reference is made to the Survey Guidelines under the Harmonized System of Survey and Certification, 2017 (resolution A.1120(30)), as may be amended.

**1.7**     The PSCO should bear in mind that a ship may be equipped over and above the requirements of Annex II of MARPOL. If such equipment is malfunctioning the flag State should be informed. This alone, however, should not cause a ship to be detained unless the malfunction presents an unreasonable threat of harm to the marine environment.

## 2     Ships of non-Parties to the Convention

**2.1**     As this category of ship is not provided with a COF or NLS Certificate as required by Annex II of MARPOL, the PSCO should be satisfied with regard to the construction and equipment standards relevant to the ship on the basis of the requirements set out in Annex II of MARPOL and the Standards for Procedures and Arrangements.

**2.2**     In all other respects the PSCO should be guided by the procedures for ships referred to in section 1 above (i.e. ships required to hold a Certificate).

**2.3**     If the ship has some form of certification other than the required Certificate, the PSCO may take the form and content of this document into account in the evaluation of that ship. Such a form of certification, however, is only of value to the PSCO if the ship has been provided with a P and A Manual.

## 3     Control

In exercising the control functions, the PSCO should use professional judgement to determine whether to detain the ship until any noted deficiencies are rectified or to allow it to sail with certain deficiencies which do not pose an unreasonable threat of harm to the marine environment. In doing this, the PSCO should be guided by the principle that the requirements contained in Annex II of MARPOL, in respect of construction and equipment and the operation of ships, are essential for the protection of the marine environment and that departure from these requirements could constitute an unreasonable threat of harm to the marine environment.

## Part 2
*Contravention of discharge provisions*

**1**     With illegal discharges, past experience has shown that information furnished to the flag State is often inadequate to enable the flag State to cause proceedings to be brought in respect of the alleged violation of the discharge requirements. This appendix is intended to identify information which will be needed by a flag State for the prosecution of violations of the discharge provisions under Annex II of MARPOL.

**2**     It is recommended that in preparing a port State report on deficiencies, where contravention of the discharge requirements is involved, the authorities of a coastal or port State should be guided by the itemized list of possible evidence as shown in part 3 of this appendix. It should be borne in mind in this connection that:

> **.1**     the report aims to provide the optimal collation of obtainable data; however, even if all the information cannot be provided, as much information as possible should be submitted;

> **.2**     it is important for all the information included in the report to be supported by facts which, when considered as a whole, would lead the port or coastal State to believe a contravention has occurred; and

> **.3**     the discharge may have been oil, in which case part 2 to appendix 3 of this resolution applies (Guidelines for Investigation and Inspections carried out under Annex I of MARPOL).

**3**     In addition to the port State report on deficiencies, a report should be completed by a port or coastal State, on the basis of the itemized list of possible evidence. It is important that these reports are supplemented by documents such as:

> **.1**     a statement by the observer of the pollution; in addition to the information required under section 1 of part 3 of this appendix, the statement should include considerations which have led the observer to conclude that none of any other possible pollution sources is in fact the source;

.2     statements concerning the sampling procedures both of the slick and on board; these include location of and time when samples were taken, identity of person(s) taking the samples and receipts identifying the persons having custody and receiving transfer of the samples;

.3     reports of analyses of samples taken of the slick and on board; the reports should include the results of the analyses, a description of the method employed, reference to or copies of scientific documentation attesting to the accuracy and validity of the method employed and names of persons performing the analyses and their experience;

.4     a statement by the PSCO on board together with the PSCO's rank and organization;

.5     statements by persons being questioned;

.6     statements by witnesses;

.7     photographs of the slick; and

.8     copies of relevant pages of the Cargo Record Book, logbooks, discharge recordings, etc.

**4**     All observations, photographs and documentation should be supported by a signed verification of their authenticity. All certifications, authentications or verifications shall be executed in accordance with the laws of the State which prepares them. All statements should be signed and dated by the person making the statement and, if possible, by a witness to the signing. The names of the persons signing statements should be printed in legible script above or below the signature.

**5**     The report referred to in paragraphs 2 and 3 should be sent to the flag State. If the coastal State observing the pollution and the port State carrying out the investigation on board are not the same, the State carrying out the latter investigation should also send a copy of its findings to the State observing the pollution and requesting the investigation.

# Part 3
*Itemized list of possible evidence on alleged contravention of the MARPOL Annex II discharge provisions*

## 1     Action on sighting pollution

### 1.1     Particulars of ship or ships suspected of contravention

.1     Name of ship and IMO Number

.2     Reasons for suspecting the ship

.3     Date and time (UTC) of observation or identification

.4     Position of ship

.5     Flag and port of registry

.6     Type, size (estimated tonnage) and other descriptive data (e.g. superstructure, colour and funnel mark)

.7     Draught condition (loaded or in ballast)

.8     Approximate course and speed

.9     Position of slick in relation to ship (e.g. astern, port, starboard)

.10     Part of the ship from which discharge was seen emanating

.11     Whether discharge ceased when ship was observed or contacted by radio

## 1.2 Particulars of slick

**.1** Date and time (UTC) of observation if different from item 1.1.3

**.2** Position of slick in longitude and latitude if different from item 1.1.4

**.3** Approximate distance in nautical miles from the nearest land

**.4** Depth of water according to sea chart

**.5** Approximate overall dimension of slick (length, width and percentage thereof covered)

**.6** Physical description of slick (direction and form, e.g. continuous, in patches or in windrows)

**.7** Colour of slick

**.8** Sky conditions (bright sunshine, overcast, etc.), lightfall and visibility (km) at the time of observation

**.9** Sea state

**.10** Direction and speed of surface wind

**.11** Direction and speed of current

## 1.3 Identification of the observer(s)

**.1** Name of the observer

**.2** Organization with which observer is affiliated (if any)

**.3** Observer's status within the organization

**.4** Observation made from aircraft, ship, shore or otherwise

**.5** Name or identity of ship or aircraft from which the observation was made

**.6** Specific location of ship, aircraft, place on shore or otherwise from which observation was made

**.7** Activity engaged in by observer when observation was made, for example: patrol, voyage, flight (en route from ... to ...)

## 1.4 Method of observation and documentation

**.1** Visual

**.2** Conventional photographs

**.3** Remote sensing records and/or remote sensing photographs

**.4** Samples taken from slick

**.5** Any other form of observation (specify)

**Note:** A photograph of the discharge should preferably be in colour. The best results may be obtained with the following three photographs:

– details of the slick taken almost vertically down from an altitude of less than 300 metres with the sun behind the photographer;

– an overall view of the ship and slick showing a substance emanating from the particular ship; and

– details of the ship for the purposes of identification

## 1.5    Other information if radio contact can be established

.1    Master informed of pollution

.2    Explanation of master

.3    Ship's last port of call

.4    Ship's next port of call

.5    Name of ship's master and owner

.6    Ship's call sign

# 2    Investigation on board

## 2.1    Inspection of the Certificate (COF or NLS Certificate)

.1    Name of ship and IMO Number

.2    Distinctive number or letters

.3    Port of registry

.4    Type of ship

.5    Date and place of issue

.6    Date and place of endorsement

.7    List of Annex II substances the ship is certified to carry

.8    Limitation as to tanks in which these substances may be carried

## 2.2    Inspection of P and A Manual

.1    Ship equipped with an efficient stripping system

.2    Residue quantities established at survey

## 2.3    Inspection of Cargo Record Book (CRB)

Copy sufficient pages of the CRB to cover a full loading/unloading/ballasting and tank cleaning cycle of the ship. Also copy the tank diagram.

## 2.4    Inspection of logbook

.1    Last port, date of departure, draught forward and aft

.2    Current port, date of arrival, draught forward and aft

.3    Ship's position at or near the time the incident was reported

.4    Spot check if times entered in the Cargo Record Book in respect of discharges correspond with sufficient distance from the nearest land, the required ship's speed and with sufficient water depth

## 2.5    Inspection of other documentation on board

**Other documentation relevant for evidence (if necessary, make copies) such as:**

–    cargo documents of cargo presently or recently carried, together with relevant information on required unloading temperature, viscosity and/or melting point

–    records of temperature of substances during unloading

–    records of monitoring equipment if fitted

## 2.6    Inspection of ship

    **.1**    Ship's equipment in accordance with the P and A Manual

    **.2**    Samples taken; state location on board

    **.3**    Sources of considerable leakage

    **.4**    Cargo residues on surface of segregated or dedicated clean ballast

    **.5**    Condition of pump-room bilges

    **.6**    Condition of monitoring system

    **.7**    Slop tank contents (estimate quantity of water and residues)

## 2.7    Statements of persons concerned if the CRB has not been properly completed, information on the following questions may be pertinent

    **.1**    Was there a discharge (accidental or intentional) at the time indicated on the incident report?

    **.2**    Which tanks are going to be loaded in the port?

    **.3**    Which tanks needed cleaning at sea? Had the tanks been prewashed?

    **.4**    When and where were these cleaned?

    **.5**    Residues of which substances were involved?

    **.6**    What was done with the tank washing slops?

    **.7**    Was the slop tank, or cargo tank used as a slop tank, discharged at sea?

    **.8**    When and where was the discharge effected?

    **.9**    What are the contents of the slop tank or cargo tank used as slop tank?

    **.10**    Which tanks contained the dirty ballast during the ballast voyage (if ship arrived in ballast)?

    **.11**    Which tanks contained the clean ballast during the ballast voyage (if ship arrived in ballast)?

    **.12**    Details of the present voyage of the ship (previous ports, next ports, trade)

    **.13**    Difficulties experienced with discharge to shore reception facilities

    **.14**    Difficulties experienced with efficient stripping operations

    **.15**    Which tanks are clean or dirty on arrival?

    **.16**    Repairs carried out or envisaged in cargo tanks

### Miscellaneous information

    **.17**    Comments in respect of condition of ship's equipment

    **.18**    Comments in respect of pollution report

    **.19**    Other comments

## 3    Investigation ashore

## 3.1    Analyses of samples

Indicate method and results of the samples' analyses.

## 3.2    Further information

Additional information on the ship, obtained from terminal staff, tank cleaning contractors or shore reception facilities may be pertinent.

**Note:**   Any information under this heading is, if practicable, to be corroborated by documentation such as signed statements, invoices, receipts, etc.

## 3.3    Information from previous unloading port terminal

.1    Confirmation that the ship was unloaded, stripped or prewashed in accordance with its P and A Manual

.2    The nature of difficulties if any

.3    Restrictions by authorities under which the ship was permitted to sail

.4    Restrictions in respect of shore reception facilities

## 4    Information not covered by the foregoing

## 5    Conclusion

.1    Summing up of the investigator's conclusions

.2    Indication of applicable provisions of Annex II of MARPOL which the ship is suspected of having contravened

.3    Did the results of the investigation warrant the filing of a deficiency report?

# Part 4
*Procedures for inspection of unloading, stripping and prewashing operations (mainly in unloading ports)*

## 1    Introduction

The PSCO or the surveyor authorized by the Administration exercising control in accordance with regulation 16 of MARPOL Annex II should be thoroughly acquainted with Annex II of MARPOL and the custom of the port as of relevance to cargo handling, tank washing, cleaning berths, prohibition of lighters alongside, etc.

## 2    Documentation

The documentation required for the inspection referred to in this appendix consists of:

.1    COF or NLS Certificate;

.2    cargo plan and shipping document;

.3    Procedures and Arrangements (P and A) Manual; and

.4    Cargo Record Book.

## 3    Information by ship's staff

3.1    Of relevance to the PSCO or the surveyor appointed or authorized by the Administration is the following:

.1    the intended loading and unloading programme of the ship;

.2     whether unloading and stripping operations can be effected in accordance with the P and A Manual and if not the reason why it cannot be done;

.3     the constraints, if any, under which the efficient stripping system operates (i.e. back pressure, ambient air temperature, malfunctioning, etc.); and

.4     whether the ship requests an exemption from the prewashing and the discharge of residues in the unloading port.

**3.2**     When tank washing is required without the use of water the PSCO or the surveyor appointed or authorized by the Administration is to be informed about the tank washing procedure and disposal of residues.

**3.3**     When the Cargo Record Book is not up to date, any information on prewash and residue disposal operations outstanding should be supplied.

# 4     Information from terminal staff

Terminal staff should supply information on limitations imposed upon the ship in respect of back pressure and/or reception facilities.

# 5     Control

**5.1**     On boarding and introduction to the master or responsible ship officers, the PSCO or the surveyor appointed or authorized by the Administration should examine the necessary documentation.

**5.2**     The documentation may be used to establish the following:

.1     noxious liquid substances to be unloaded, their categories and stowage (cargo plan, P and A Manual);

.2     details of efficient stripping system, if fitted (P and A Manual);

.3     tanks which require prewashing with disposal of tank washings to reception facilities (shipping document and cargo temperature);

.4     tanks which require prewashing with disposal of tank washings either to reception facilities or into the sea (P and A Manual, shipping document and cargo temperature);

.5     prewash operations and/or residue disposal operations outstanding (Cargo Record Book); and

.6     tanks which may not be washed with water due to the nature of substances involved (P and A Manual).

**5.3**     In respect of the prewash operations referred to under paragraph 5.2, the following information is of relevance (P and A Manual):

.1     pressure required for tank washing machines;

.2     duration of one cycle of the tank washing machine and quantity of water used;

.3     washing programmes for the substances involved;

.4     required temperature of washing water; and

.5     special procedures.

**5.4**    The PSCO or the surveyor authorized by the Administration, in accordance with regulation 16 of MARPOL Annex II, should ascertain that unloading, stripping and/or prewash operations are carried out in conformance with the information obtained in accordance with paragraph 2 (Documentation) of this Part. If this cannot be achieved, alternative measures should be taken to ensure that the ship does not proceed to sea with more than the quantities of residue specified in regulation 12 of MARPOL Annex II, as applicable. If the residue quantities cannot be reduced by alternative measures the PSCO or the surveyor appointed or authorized by the Administration should inform the port State Administration.

**5.5**    Care should be taken to ensure that cargo hoses and piping systems of the terminal are not drained back to the ship.

**5.6**    If a ship is exempted from certain pumping efficiency requirements under regulation 4.4 of MARPOL Annex II or requests an exemption from certain stripping or prewashing procedures under regulation 13.4 of MARPOL Annex II the conditions for such exemption set out in the said regulations should be observed. These concern:

.1    regulations 4.2 and 4.3: the ship is constructed before 1 July 1986 and is exempted from the requirement for reducing its residue quantities to specified limits of regulation 12 (i.e. category X or Y substances 300 litres and category Z substances 900 litres); this is subject to the conditions of regulation 4.3 that whenever a cargo tank is to be washed or ballasted, a prewash is required with disposal of prewash slops to shore reception facilities; the COF or NLS Certificate should have been endorsed to the effect that the ship is solely engaged in restricted voyages;

.2    regulation 4.4: the ship is never required to ballast its cargo tanks and tank washing is only required for repair or dry-docking; the COF or NLS Certificate should indicate the particulars of the exemption; each cargo tank should be certified for the carriage of only one named substance;

.3    regulation 13.4.1: cargo tanks will not be washed or ballasted prior to the next loading;

.4    regulation 13.4.2: cargo tanks will be washed and prewash slops will be discharged to reception facilities in another port; it should be confirmed in writing that an adequate reception facility is available at that port for such purpose; and

.5    regulation 13.4.3: the cargo residues can be removed by ventilation.

**5.7**    The PSCO or the surveyor appointed or authorized by the Administration must endorse the Cargo Record Book under section J whenever an exemption under regulation 13.4 referred to in paragraph 5.6 above has been granted, or whenever a tank having unloaded category X substances has been prewashed in accordance with the P and A Manual.

**5.8**    Alternatively, for category X substances, regulation 13.6.1.1 of MARPOL Annex II, residual concentration should be measured by the procedures which each port State authorizes. In this case the PSCO or the surveyor authorized by the Administration must endorse in the Cargo Record Book under section K whenever the required residual concentration has been achieved.

**5.9**    In addition to paragraph 5.7 above, the PSCO or the surveyor authorized by the Administration shall endorse the Cargo Record Book whenever the unloading, stripping or prewash of category Y and Z substances, in accordance with the P and A Manual, has actually been witnessed.

# Appendix 5
*Guidelines for discharge requirements under*
*Annexes I and II of MARPOL*

## 1    Introduction

**1.1**    Regulations 15 and 34 of MARPOL Annex I prohibit the discharge into the sea of oil and regulation 13 of Annex II prohibits the discharge into the sea of noxious liquid substances except under precisely defined conditions. A record of these operations shall be completed, where appropriate, in the form of an Oil or Cargo Record Book as applicable and shall be kept in such a place as to be readily available for inspection at all reasonable times.

**1.2**    The regulations referred to above provide that whenever visible traces of oil are observed on or below the surface of the water in the immediate vicinity of a ship or of its wake, a Party should, to the extent that it is reasonably able to do so, promptly investigate the facts bearing on the issue of whether or not there has been a violation of the discharge provisions.

**1.3**    The conditions under which noxious liquid substances are permitted to be discharged into the seas include quantity, quality and position limitations, which depend on category of substance and sea area.

**1.4**    An investigation into an alleged contravention should therefore aim to establish whether a noxious liquid substance has been discharged and whether the operations leading to that discharge were in accordance with the ship's Procedures and Arrangements Manual (P and A Manual).

**1.5**    Recognizing the likelihood that many of the violations of the discharge provisions will take place outside the immediate control and knowledge of the flag State, article 6 of MARPOL provides that Parties shall cooperate in the detection of violations and the enforcement of the provisions using all appropriate and practicable measures of detection and environmental monitoring, adequate procedures for reporting and gathering evidence. MARPOL also contains a number of more specific provisions designed to facilitate that cooperation.

**1.6**    Several sources of information about possible violations of the discharge provisions can be indicated. These include:

.1    reports by masters: article 8 and Protocol I of MARPOL require, inter alia, a ship's master to report certain incidents involving the discharge or the probability of a discharge of oil or oily mixtures, or noxious liquid substances or mixtures containing such substances;

.2    reports by official bodies: article 8 of MARPOL requires furthermore that a Party issue instructions to its maritime inspection vessels and aircraft and to other appropriate services to report to its authorities incidents involving the discharge or the probability of a discharge of oil or oily mixtures, or noxious liquid substances or mixtures containing such substances;

.3    reports by other Parties: article 6 of MARPOL provides that a Party may request another Party to inspect a ship; the Party making the request shall supply sufficient evidence that the ship has discharged oil or oily mixtures, noxious liquid substances or mixtures containing such substances, or that the ship has departed from the unloading port with residues of noxious liquid substances in excess of those permitted to be discharged into the sea; and

.4    reports by others: it is not possible to list exhaustively all sources of information concerning alleged contravention of the discharge provisions; Parties should take all circumstances into account when deciding upon investigating such reports.

**1.7**    Action which can be taken by States other than the flag or port States that have information on discharge violations (hereinafter referred to as coastal States):

   **.1**    coastal States, Parties to MARPOL, upon receiving a report of pollution by oil or noxious liquid substances allegedly caused by a ship, may investigate the matter and collect such evidence as can be collected; for details of the desired evidence reference is made to appendices 3 and 4;

   **.2**    if the investigation referred to under subparagraph .1 above discloses that the next port of call of the ship in question lies within its jurisdiction, the coastal State should also take port State action as set out in paragraphs 2.1 to 2.6 below;

   **.3**    if the investigation referred to in subparagraph .1 above discloses that the next port of call of the ship in question lies within the jurisdiction of another Party, then the coastal State should in appropriate cases furnish the evidence to that other Party and request that Party to take port State action in accordance with paragraphs 2.1 to 2.6 below; and

   **.4**    in either case referred to in subparagraphs .2 and .3 above and if the next port of call of the ship in question cannot be ascertained, the coastal State shall inform the flag State of the incident and of the evidence obtained.

## 2    Port State action

**2.1**    Parties shall appoint or authorize officers to carry out investigations for the purpose of verifying whether a ship has discharged oil or noxious liquid substances in violation of the provisions of MARPOL.

**2.2**    Parties may undertake such investigations on the basis of reports received from sources indicated in paragraph 1.6 above.

**2.3**    These investigations should be directed toward the gathering of sufficient evidence to establish whether the ship has violated the discharge requirements. Guidelines for the optimal collation of evidence are given in appendices 3 and 4.

**2.4**    If the investigations provide evidence that a violation of the discharge requirements took place within the jurisdiction of the port State, that port State shall either cause proceedings to be taken in accordance with its law, or furnish to the flag State all information and evidence in its possession about the alleged violation. When the port State causes proceedings to be taken, it shall inform the flag State.

**2.5**    Details of the report to be submitted to the flag State are set out in appendix 16.

**2.6**    The investigation might provide evidence that pollution was caused through damage to the ship or its equipment. This might indicate that a ship is not guilty of a violation of the discharge requirements of Annex I or Annex II of MARPOL provided that:

   **.1**    all reasonable precautions have been taken after the occurrence of the damage or discovery of the discharge for the purpose of preventing or minimizing the discharge; and

   **.2**    the owner or the master did not act either with intent to cause damage or recklessly and with knowledge that damage would probably result.

**2.7**    However, action by the port State as set out in chapter 3 of these Procedures may be called for.

## 3    Inspection of crude oil washing (COW) operations

**3.1**    Regulations 18, 33 and 35 of MARPOL Annex I, inter alia, require that crude oil washing of cargo tanks be performed on certain categories of crude carriers. A sufficient number of tanks shall be washed in order that ballast water is put only in cargo tanks which have been crude oil washed. The remaining cargo tanks shall be washed on a rotational basis for sludge control.

**3.2**    Port State Authorities may carry out inspections to ensure that crude oil washing is performed by all crude carriers either required to have a COW system or where the owner or operator chooses to install a COW system in order to comply with regulation 18 of MARPOL Annex I. In addition compliance should be ensured with the operational requirements set out in the Revised specifications for the design, operation and control of crude oil washing systems (resolution A.446(XI), as amended). This can best be done in the ports where the cargo is unloaded.

**3.3**    Parties should be aware that the inspection referred to in paragraph 3.2 may also lead to the identification of a pollution risk, necessitating additional action by the port State as set out in chapter 3 of these Procedures.

**3.4**    Detailed guidelines for in-port inspections of crude oil washing procedures have been approved and published by IMO (Crude Oil Washing Systems, revised edition, 2000) and are set out in part 4 to appendix 3.

## 4    Inspection of unloading, stripping and prewash operations

**4.1**    Regulation 16 of MARPOL Annex II requires Parties to MARPOL to appoint or authorize surveyors for the purpose of implementing the regulation.

**4.2**    The provisions of regulation 16 are aimed at ensuring in principle that a ship having unloaded, to the maximum possible extent, noxious liquid substances of category X, Y or Z, proceeds to sea only if residues of such substances have been reduced to such quantities as may be discharged into the sea.

**4.3**    Compliance with these provisions is in principle ensured in the case of categories X, Y and Z substances through the application of a prewash in the unloading port and the discharge of prewash residue water mixtures to reception facilities, except that in the case of non-solidifying and low viscosity categories Y and Z substances, requirements for the efficient stripping of a tank to negligible quantities apply in lieu of the application of a prewash. Alternatively for a number of substances ventilation procedures may be employed for removing cargo residues from a tank.

**4.4**    Regulation 16.6 permits the Government of the receiving Party to exempt a ship proceeding to a port or terminal under the jurisdiction of another Party from the requirement to prewash cargo tanks and discharge residue/water mixtures to a reception facility.

**4.5**    Existing chemical tankers engaged on restricted voyages may by virtue of regulation 4.3 of MARPOL Annex II be exempted from the quantity limitation requirements of regulations 12.1 to 12.3. If a cargo tank is to be ballasted or washed, a prewash is required after unloading category Y or Z substances and prewash residue water mixtures must be discharged to shore reception facilities. The exemption should be indicated on the certificate.

**4.6**    A ship whose constructional and operational features are such that ballasting of cargo tanks is not required and cargo tank washing is only required for repairs or dry-docking may by virtue of regulation 4.4 be exempted from the provisions of regulation 12 of MARPOL Annex II provided that all conditions mentioned in regulation 4.4 are complied with. Consequentially, the certificate of the ship should indicate that each cargo tank is only certified for the carriage of one named substance. It should also indicate the particulars of the exemption granted by the Administration in respect of pumping, piping and discharge arrangements.

**4.7**    Detailed instructions on efficient stripping and prewash procedures are included in a ship's Procedures and Arrangements Manual. The Manual also contains alternative procedures to be followed in case of equipment failure.

**4.8**    Parties should be aware that the inspection referred to in paragraphs 1.3 and 1.4 above may lead to the identification of a pollution risk or of a contravention of the discharge provisions, necessitating port State action as set out in chapter 3 of these Procedures.

**4.9**    For details in respect of inspections under this section reference is made to appendix 4.

# Appendix 6
*Guidelines for more detailed inspections of ship structural and equipment requirements*

## 1      Introduction

If the PSCO from general impressions or observations on board has clear grounds for believing that the ship might be substandard, the PSCO should proceed to a more detailed inspection, taking the following considerations into account.

## 2      Structure

**2.1**      The PSCO's impression of hull maintenance and the general state on deck, the condition of such items as ladderways, guard rails, pipe coverings and areas of corrosion or pitting should influence the PSCO's decision as to whether it is necessary to make the fullest possible examination of the structure with the ship afloat. Significant areas of damage or corrosion, or pitting of plating and associated stiffening in decks and hull affecting seaworthiness or strength to take local loads, may justify detention. It may be necessary for the underwater portion of the ship to be checked. In reaching a decision, the PSCO should have regard to the seaworthiness and not the age of the ship, making an allowance for fair wear and tear over the minimum acceptable scantlings. Damage not affecting seaworthiness will not constitute grounds for judging that a ship should be detained, nor will damage that has been temporarily but effectively repaired for a voyage to a port for permanent repairs. However, in this assessment of the effect of damage, the PSCO should have regard to the location of crew accommodation and whether the damage substantially affects its habitability.

**2.2**      The PSCO should pay particular attention to the structural integrity and seaworthiness of bulk carriers and oil tankers and note that these ships must undergo the enhanced programme of inspection during surveys under the provision of regulation XI-1/2 of SOLAS.

**2.3**      The PSCO's assessment of the safety of the structure of those ships should be based on the Survey Report File carried on board. This file should contain reports of structural surveys, condition evaluation reports (translated into English and endorsed by or on behalf of the Administration), thickness measurement reports and a survey planning document. The PSCO should note that there may be a short delay in the update of the Survey Report File following survey. Where there is doubt that the required survey has taken place, the PSCO should seek confirmation from the recognized organization.

**2.4**      If the Survey Report File necessitates a more detailed inspection of the structure of the ship or if no such report is carried, special attention should be given by the PSCO, as appropriate, to hull structure, piping systems in way of cargo tanks or holds, pump-rooms, cofferdams, pipe tunnels, void spaces within the cargo area, and ballast tanks.

**2.5**      For bulk carriers, PSCOs should inspect holds' main structure for any obviously unauthorized repairs. For bulk carriers the port State control officer should verify that the bulk carrier booklet has been endorsed, the water level alarms in cargo holds are fitted, and where applicable, that any restrictions imposed on the carriage of solid bulk cargoes have been recorded in the booklet and the bulk carrier loading triangle is permanently marked.

## 3      Machinery spaces

**3.1**      The PSCO should assess the condition of the machinery and of the electrical installations such that they are capable of providing sufficient continuous power for propulsion and for auxiliary services.

**3.2**     During inspection of the machinery spaces, the PSCO should form an impression of the standard of maintenance. Frayed, disconnected or inoperative quick-closing valve wires, disconnected or inoperative extended control rods or machinery trip mechanisms, missing valve hand wheels, evidence of chronic steam, water and oil leaks, dirty tank tops and bilges or extensive corrosion of machinery foundations are pointers to an unsatisfactory organization of the systems' maintenance. A large number of temporary repairs, including pipe clips or cement boxes, will indicate reluctance to make permanent repairs.

**3.3**     While it is not possible to determine the condition of the machinery without performance trials, general deficiencies, such as leaking pump glands, dirty water gauge glasses, inoperable pressure gauges, rusted relief valves, inoperative or disconnected safety or control devices, evidence of repeated operation of diesel engine scavenge belt or crankcase relief valves, malfunctioning or inoperative automatic equipment and alarm systems, and leaking boiler casings or uptakes, would warrant inspection of the engine-room logbook and investigation into the record of machinery failures and accidents and a request for running tests of machinery.

**3.4**     If one electrical generator is out of commission, the PSCO should investigate whether power is available to maintain essential and emergency services and should conduct tests.

**3.5**     If evidence of neglect becomes evident, the PSCO should extend the scope of an investigation to include, for example, tests on the main and auxiliary steering gear arrangements, overspeed trips, circuit breakers, etc.

**3.6**     It must be stressed that while detection of one or more of the above deficiencies would afford guidance to a substandard condition, the actual combination is a matter for professional judgement in each case.

# 4     Conditions of assignment of load lines

It may be that the PSCO has concluded that a hull inspection is unnecessary but, if dissatisfied on the basis of observations on deck, with items such as defective hatch closing arrangements, corroded air pipes and vent coamings, the PSCO should examine closely the conditions of assignment of load lines, paying particular attention to closing appliances, means of freeing water from the deck and arrangements concerned with the protection of the crew.

# 5     Life-saving appliances

**5.1**     The effectiveness of life-saving appliances depends heavily on good maintenance by the crew and their use in regular drills. The lapse of time since the last survey for a Safety Equipment Certificate can be a significant factor in the degree of deterioration of equipment if it has not been subject to regular inspection by the crew. Apart from failure to carry equipment required by a convention or obvious defects such as holed lifeboats, the PSCO should look for signs of disuse of, obstructions to, or defects with survival craft launching and recovery equipment, which may include paint accumulation, seizing of pivot points, absence of greasing, condition of blocks and falls, condition of lifeboat lifting hook attachment to the lifeboat hull and improper lashing or stowing of deck cargo.

**5.2**     Should such signs be evident, the PSCO would be justified in making a detailed inspection of all life-saving appliances. Such an examination might include the lowering of survival craft, a check on the servicing of liferafts, the number and condition of lifejackets and lifebuoys and ensuring that the pyrotechnics are still within their period of validity. It would not normally be as detailed as that for a renewal of the Safety Equipment Certificate and would concentrate on essentials for safe abandonment of the ship, but in an extreme case could progress to a full Safety Equipment Certificate inspection. The provision and functioning of effective overside lighting, means of alerting the crew and passengers and provision of illuminated routes to assembly points and embarkation positions should be given importance in the inspection.

# 6     Fire safety

**6.1**     Ships in general: The poor condition of fire and wash deck lines and hydrants and the possible absence of fire hoses and extinguishers in accommodation spaces might be a guide to a need for a close inspection of all fire safety equipment. In addition to compliance with convention requirements, the PSCO should look for evidence of a higher than normal fire risk; this might be brought about by a poor standard of cleanliness in the machinery space, which together with significant deficiencies of fixed or portable fire-extinguishing equipment could lead to a judgement of the ship being substandard. Queries on the method of structural protection should be addressed to the flag Administration and the PSCO should generally confine the inspection to the effectiveness of the arrangements provided.

**6.2** Passenger ships: The PSCO should initially form an opinion of the need for inspection of the fire safety arrangements on the basis of consideration of the ship under the previous headings and, in particular, that dealing with fire safety equipment. If the PSCO considers that a more detailed inspection of fire safety arrangements is necessary, the PSCO should examine the fire control plan on board in order to obtain a general picture of the fire safety measures provided in the ship and consider their compliance with convention requirements for the year of build. Queries on the method of structural protection should be addressed to the flag Administration and the PSCO should generally confine the inspection to the effectiveness of the arrangements provided.

**6.3** The spread of fire could be accelerated if fire doors are not readily operable. The PSCO should inspect for the operability and securing arrangements of those doors in the main zone bulkheads and stairway enclosures and in boundaries of high fire risk spaces, such as main machinery rooms and galleys, giving particular attention to those retained in the open position. Attention should also be given to main vertical zones which may have been compromised through new construction. An additional hazard in the event of fire is the spread of smoke through ventilation systems. Spot checks might be made on dampers and smoke flaps to ascertain the standard of operability. The PSCO should also ensure that ventilation fans can be stopped from the master controls and that means are available for closing main inlets and outlets of ventilation systems.

**6.4** Attention should be given to the effectiveness of escape routes by ensuring that vital doors are not maintained locked and that alleyways and stairways are not obstructed. Regarding the minimum width of external escape routes, the arrangements approved by the flag Administrations should be accepted.

**6.5** The arrangements for the location of manually operated call points as approved by the flag Administrations should be accepted.

# 7 Regulations for preventing collisions at sea

A vital aspect of ensuring safety of life at sea is full compliance with the collision regulations. Based on observations on deck, the PSCO should consider the need for close inspection of lanterns and their screening and means of making sound and distress signals.

# 8 Cargo Ship Safety Construction Certificate

The general condition of the ship may lead the PSCO to consider matters other than those concerned with safety equipment and assignment of load lines, but nevertheless associated with the safety of the vessel, such as the effectiveness of items associated with the Cargo Ship Safety Construction Certificate, which can include pumping arrangements, means for shutting off air and oil supplies in the event of fire, alarm systems and emergency power supplies.

# 9 Cargo Ship Safety Radio Certificates

The validity of the Cargo Ship Safety Radio Certificates and associated Record of Equipment (Form R) may be accepted as proof of the provision and effectiveness of its associated equipment, but the PSCO should ensure that appropriate certificated personnel are carried for its operation and for listening periods. Requirements for maintenance of radio equipment are contained in SOLAS regulation IV/15. The radio log or radio records should be examined. Where considered necessary, operational checks may be carried out.

## 10    Means of access to ship

**10.1**    Prior to boarding a ship, the PSCO should assess the means of embarkation on and disembarkation from the ship. The PSCO should be guided by SOLAS regulation II-1/3-9 noting its application to ships constructed on or after 1 January 2010 but also noting that paragraph 3 of this regulation applies to all ships and requires that:

.1    the means of embarkation and disembarkation shall be inspected and maintained in suitable condition for their intended purpose, taking into account any restrictions related to safe loading; and

.2    all wires used to support the means of embarkation and disembarkation shall be maintained as specified in SOLAS regulation III/20.4.

**10.2**    In regard to the maintenance of the means of embarkation and disembarkation, the PSCO should refer to the Guidelines for construction, installation, maintenance and inspection/survey of means of embarkation and disembarkation (MSC.1/Circ.1331).

**10.3**    During the inspection, the PSCO should also ensure that the pilot transfer arrangements comply with SOLAS regulation V/23 and the Unified interpretation of SOLAS regulation V/23 (MSC.1/Circ.1375/Rev.1 and MSC.1/Circ.1495/Rev.1).

## 11    Equipment in excess of convention or flag State requirements

Equipment on board which is expected to be relied on in situations affecting safety or pollution prevention must be in operating condition. If such equipment is inoperative and is in excess of the equipment required by an appropriate convention and/or the flag State, it should be repaired, removed or, if removal is not practicable, clearly marked as inoperative and secured.

# Appendix 7
*Guidelines for control of operational requirements*

## 1    Introduction

**1.1**    When, during a port State control inspection, the PSCO has clear grounds according to section 2.4 of the present Procedures, the following on-board operational procedures may be checked in accordance with this resolution.

**1.2**    However, in exercising controls recommended in these guidelines, the PSCO should not include any operational tests or impose physical demands which, in the judgement of the master, could jeopardize the safety of the ship, crew, passengers, control officers or cargo. Prior to requiring any practical operational control, the PSCO should review training and drill records and should inspect, as appropriate, the associated safety equipment and its maintenance records. For example, an enclosed space entry drill may be sufficiently verified without an actual enclosed space entry by verifying drill records, maintenance records, physical inspection and physical demonstrations by crew of breathing apparatus, safety harnesses and atmosphere testing instruments.

**1.3**    When carrying out operational control, the PSCO should ensure, as far as possible, no interference with normal shipboard operations, such as loading and unloading of cargo and ballasting, which is carried out under the responsibility of the master, nor should the PSCO require demonstration of operational aspects which would unnecessarily delay the ship.

**1.4**    Having assessed the extent to which operational requirements are complied with, the PSCO then has to exercise professional judgement to determine whether the operational proficiency of the crew as a whole is of a sufficient level to allow the ship to sail without danger to the ship or persons on board, or presenting an unreasonable threat of harm to the marine environment.

**1.5**    When assessing the crew's ability to conduct an operational drill, the mandatory minimum requirements for familiarization and basic safety training for seafarers, as stated in the STCW Convention 1978, as amended, shall be used as a benchmark.

## 2    Muster list

**2.1**    The PSCO may determine if the crew members are aware of their duties indicated in the muster list.

**2.2**    The PSCO may ensure that muster lists are exhibited in conspicuous places throughout the ship, including the navigational bridge, the engine-room and the crew accommodation spaces. When determining if the muster list is in accordance with the regulations, the PSCO may verify whether:

   .1    the muster list shows the duties assigned to the different members of the crew;

   .2    the muster list specifies which officers are assigned to ensure that life-saving and fire appliances are maintained in good condition and are ready for immediate use;

   .3    the muster list specifies the substitutes for key persons who may become disabled, taking into account that different emergencies may call for different actions;

   .4    the muster list shows the duties assigned to crew members in relation to passengers in case of emergency; and

   .5    the format of the muster list used on passenger ships is approved and is drawn up in the language or languages required by the ship's flag State and in the English language.

**2.3**    To determine whether the muster list is up to date, the PSCO may require an up-to-date crew list, if available, to verify this.

**2.4**    The PSCO may determine whether the duties assigned to crew members manning the survival craft (lifeboats or liferafts) are in accordance with the regulations and verify that a deck officer or certificated person is placed in charge of each survival craft to be used. However, the Administration (of the flag State), having due regard to the nature of the voyage, the number of persons on board and the characteristics of the ship, may permit persons practised in the handling and operation of liferafts to be placed in charge of liferafts in lieu of persons qualified as above. A second-in-command shall also be nominated in the case of lifeboats.

**2.5**    The PSCO may determine whether the crew members are familiar with the duties assigned to them in the muster list and are aware of the locations where they should perform their duties.

# 3    Communication

**3.1**    The PSCO may determine if the key crew members are able to communicate with each other, and with passengers as appropriate, in such a way that the safe operation of the ship is not impaired, especially in emergency situations.

**3.2**    The PSCO may ask the master which languages are used as the working languages and may verify whether the language has been recorded in the logbook.

**3.3**    The PSCO may ensure that the key crew members are able to understand each other during the inspection or drills. The crew members assigned to assist passengers should be able to give the necessary information to the passengers in case of an emergency.

# 4    Search and Rescue Plan

For passenger ships, the PSCO may verify that there is on board an approved plan for cooperation with appropriate search and rescue services in event of an emergency.

# 5    Fire and abandon ship drills

**5.1**    The PSCO witnessing a fire and abandon ship drill should ensure that the crew members are familiar with their duties and the proper use of the ship's installations and equipment.

**5.2**    When setting a drill scenario, witnessing the drill and finally assessing the standard of the drill, it is important to emphasize that the PSCO is not looking for an exceptional drill, particularly on cargo ships. The main points for the PSCO to be satisfied are:

   **.1**    In the event of a shipboard emergency, can the crew organize themselves into an effective team to tackle the emergency?

   **.2**    Can the crew communicate effectively?

   **.3**    Is the master in control and is information flowing to/from the command centre?

   **.4**    In the event of the situation getting out of hand can the crew safely abandon the ship?

**5.3**    It is important that when setting the scenario the PSCO clearly explains to the master exactly what is required and expected during the drill, bearing in mind there may be language difficulties. PSCOs should not be intimidating, not interfere during the drill nor offer advice. The PSCO should stand back and observe only, making appropriate notes. It is important to emphasize that the PSCO's role is not to teach or train but to witness.

**5.4**    Drills should be carried out at a safe speed. PSCOs should not expect to see operational drills conducted in real time. During drills, care should be taken to ensure that everybody familiarizes themselves with their duties and with the equipment. If necessary, drills should be stopped if the PSCO considers that the crew are carrying out unsafe practices or if there is a real emergency.

**5.5** Language difficulty between the PSCO and non-English-speaking crews can make it difficult to put across the intentions for the conduct of the exercise. Care needs to be exercised when an unsatisfactory drill takes place: this is to ensure differentiation between the crew possibly failing to understand the attending PSCO's intention and failure through lack of crew competence.

## 6 Fire drills

**6.1** The PSCO may witness a fire drill carried out by the crew assigned to these duties on the muster list. After consultation with the master of the vessel, one or more specific locations of the ship may be selected for a simulated fire. A crew member may be sent to the location(s) and activate a fire alarm system or use other means to give alarm.

**6.2** At the location the PSCO can describe the fire indication to the crew member and observe how the report of fire is relayed to the bridge or damage control centre. At this point most ships will sound the crew alarm to summon the fire-fighting parties to their stations. The PSCO should observe the fire-fighting party arriving on the scene, breaking out their equipment and fighting the simulated fire. Team leaders should be giving orders as appropriate to their crews and passing the word back to the bridge or damage control centre on the conditions. The fire-fighting crews should be observed for proper donning and use of their equipment. The PSCO should make sure that all the gear is complete. Merely mustering the crew with their gear is not acceptable. Crew response to personnel injuries can be checked by selecting a crew member as a simulated casualty. The PSCO should observe how the word is passed and the response of stretcher and medical teams. Handling a stretcher properly through narrow passageways, doors and stairways is difficult and takes practice.

**6.3** The drill should, as far as practicable, be conducted as if there were an actual emergency.

**6.4** Those crew members assigned to other duties related to a fire drill, such as the manning of the emergency generators, the $CO_2$ room, the sprinkler and emergency fire pumps, should also be involved in the drill. The PSCO may ask these crew members to explain their duties and if possible to demonstrate their familiarity.

**6.5** On passenger ships, special attention should be paid to the duties of those crew members assigned to the closing of manually operated doors and fire dampers. These closing devices should be operated by the responsible persons in the areas of the simulated fire(s) during the drill. Crew members not assigned to the fire-fighting teams are generally assigned to locations throughout the passenger accommodations to assist in passenger evacuation. These crew members should be asked to explain their duties and the meaning of the various emergency signals and asked to point out the two means of escape from the area, and where the passengers are to report. Crew members assigned to assist passengers should be able to communicate at least enough information to direct a passenger to the proper muster and embarkation stations.

## 7 Abandon ship drills

**7.1** After consultation with the master, the PSCO may require an abandon ship drill for one or more survival craft. The essence of this drill is that the survival craft are manned and operated by the crew members assigned to them on the muster list. If possible the PSCO should include the rescue boat(s) in this drill. SOLAS chapter III gives specific requirements on abandon ship training and drills, of which the following principles are particularly relevant.

**7.2** The drill should, as far as practicable, be conducted as if there were an actual emergency.

**7.3** The abandon ship drill should include:

   **.1** summoning of (passengers and) crew to the muster station(s) with the required alarm and ensuring that they are aware of the order to abandon ship as specified in the muster list;

   **.2** reporting to the stations and preparing for the duties described in the muster list;

   **.3** checking that (passengers and) crew are suitably dressed;

   **.4** checking that lifejackets are correctly donned;

.5    lowering of at least one lifeboat after the necessary preparation for launching;

.6    starting and operating the lifeboat engine;

.7    operation of the davits used for launching liferafts;

.8    a mock search and rescue of passenger trapped in their staterooms (if applicable);

.9    instructions in the use of radio life-saving appliances;

.10    testing of emergency lighting for mustering and abandonment; and

.11    if the ship is fitted with marine evacuation systems, exercising of the procedures required for the deployment of such systems up to the point immediately preceding actual deployment.

**7.4**    If the lifeboat lowered during the drill is not the rescue boat, the rescue boat should be lowered as well, taking into account that it is boarded and launched in the shortest possible time. The PSCO should ensure that crew members are familiar with the duties assigned to them during abandon ship operations and that the crew member in charge of the survival craft has complete knowledge of the operation and equipment of the survival craft. Care needs to be taken when requiring a ship to lower lifeboats. The number of persons inside the lifeboats during launching for the purpose of a drill should be at the master's discretion noting that SOLAS does not require persons in the lifeboat during lowering and recovery. The purpose of this is to reduce the risk of accidents during launching and recovery; however, this must be balanced out with the risk of embarking/disembarking the boat while it is in the water, if the boat is to be taken away and run.

**7.5**    Each survival craft should be stowed in a state of continuous readiness so that two crew members can carry out preparations for embarking and launching in less than five minutes.

**7.6**    On passenger ships, it is required that lifeboats and davit-launched liferafts are capable of being launched within a period of 30 min after all persons have been assembled with lifejackets donned.

**7.7**    On cargo ships, it is required that lifeboats and davit launched liferafts are capable of being launched within a period of 10 min.

## 8    Enclosed space entry and rescue drills

**8.1**    After consultation with the master, the PSCO may require an enclosed space entry and rescue drill. The essence of this drill is to confirm that crew members are familiar with the procedure to enter enclosed space and rescue personnel safely, can demonstrate an enclosed space entry and rescue drill and can communicate effectively when entering an enclosed space in case of planned entry and/or an emergency situation.

**8.2**    The place of the drill can be selected at an assumed enclosed space; it is not necessary to select an actual enclosed space.

**8.3**    The PSCO should check the structure of the enclosed space, the scenarios of the drills and the responsible officers listed on the master list where applicable.

**8.4**    The enclosed space entry and rescue drill should include:

.1    checking and use of personal protective equipment required for entry;

.2    checking and use of communication equipment and procedures;

.3    checking and use of instruments for measuring the atmosphere in enclosed spaces;

.4    checking use of rescue equipment and procedures; and

.5    instructions in first aid and resuscitation techniques.

## 9    Emergency steering drills

**9.1**    After consultation with the master, the PSCO may require an emergency steering drill. The essence of this drill is to confirm crew members are familiar with the procedure of emergency steering.

**9.2**     The PSCO may check the procedure and means of communication at both the navigation bridge and the steering gear room.

**9.3**     The emergency steering drills should include:

> **.1**     direct control within the steering gear compartment;
>
> **.2**     communication procedure with the navigational bridge; and
>
> **.3**     operation of alternative power supplies where applicable.

## 10     Assessment of drills

**10.1**     When witnessing a drill, the PSCO should seek:

> **.1**     confirmation that the crew follow what is required of them by the muster list;
>
> **.2**     confirmation that there are sufficient personnel assigned to the various parties to cope with the duties given to them;
>
> **.3**     confirmation that there is an effective means of communication between the party, the party leader and the bridge and that relevant information is being passed bi-directionally;
>
> **.4**     confirmation of the efficiency of the crew working as a team; this would be based on questioning of personnel and observation of their actions; the response times should be noted of the various parties in assembling at their stations; the reaction of the parties to unplanned events should also be noted;
>
> **.5**     confirmation that key members of the crew are able to understand each other;
>
> **.6**     confirmation of the efficiency of the equipment used, for example:
>
> > **.1**     that the fire alarms are audible and efficient;
> >
> > **.2**     that the fire doors close as required; and
> >
> > **.3**     that items of personal fire-fighting equipment appear well maintained; and
>
> **.7**     confirmation that the response time was considered fast enough (taking into account safety of the drill as indicated in paragraph 5.4 of this appendix), considering the size of the ship and the locations of fire, personnel and fire-fighting equipment.

**10.2**     If the PSCO determines that the crew are unfamiliar with their duties or incapable of safely operating the life-saving and fire-fighting equipment, the PSCO should halt the drill, notify the master that the drill was unsuccessful and use their professional judgement to establish the next steps, noting the likelihood that this will establish "clear grounds" for a more detailed inspection.

## 11     Damage control plan and Shipboard Oil Pollution Emergency Plan (SOPEP) or Shipboard Marine Pollution Emergency Plans (SMPEP)

**11.1**     The PSCO may determine if a damage control plan is provided on a passenger ship and whether the crew members are familiar with their duties and the proper use of the ship's installations and equipment for damage control purposes. The same applies with regard to SOPEP on all ships and SMPEP where applicable.

**11.2**     The PSCO may determine if the officers of the ship are aware of the contents of the damage control booklet which should be available to them, or of the damage control plan.

**11.3**     The officers may be asked to explain the action to be taken in various damage conditions.

**11.4**     The officers may also be asked to explain about the boundaries of the watertight compartments, the openings therein with the means of closure and position of any controls thereof and the arrangements for the correction of any list due to flooding.

**11.5**    The officers should have a sound knowledge of the effect of trim and stability of their ship in the event of damage to and consequent flooding of a compartment and counter-measures to be taken.

## 12    Fire control plan

**12.1**    The PSCO may determine if a fire control plan or booklet is provided and whether the crew members are familiar with the information given in the fire control plan or booklet.

**12.2**    The PSCO may verify that fire control plans are permanently exhibited for the guidance of the ship's officers. Alternatively, booklets containing the information of the fire control plan may be supplied to each officer, and one copy should at all times be available on board in an accessible position. Plans and booklets should be kept up to date, any alterations being recorded thereon as soon as possible.

**12.3**    The PSCO may determine that the responsible officers, especially those who are assigned to related duties on the muster list, are aware of the information provided by the fire control plan or booklet and how to act in case of a fire.

**12.4**    The PSCO may ensure that the officers in charge of the ship are familiar with the principal structural members which form part of the various fire sections and the means of access to the different compartments.

## 13    Bridge operation

**13.1**    The PSCO may determine if officers in charge of a navigational watch are familiar with bridge control and navigational equipment, changing the steering mode from automatic to manual and vice versa, and the ship's manoeuvring characteristics.

**13.2**    The officer in charge of a navigational watch should have knowledge of the location and operation of all safety and navigational equipment. Moreover, this officer should be familiar with procedures which apply to the navigation of the ship in all circumstances and should be aware of all information available.

**13.3**    The PSCO may also verify the familiarity of the officers on all the information available to them such as manoeuvring characteristics of the ship, life-saving signals, up-to-date nautical publications, checklists concerning bridge procedures, instructions, manuals, etc.

**13.4**    The Permit to Operate High-Speed Craft includes limitations of the maximum significant wave height (and wind force for hovercraft) within which the craft may operate. When carrying out inspections of HSC, PSCOs may verify by the logbook and the weather records whether these limitations have been respected. PSCOs may find that a voyage had to be completed when worse weather conditions than permitted were encountered, but a new voyage should not commence in such conditions.

**13.5**    The PSCO may verify the familiarity of the officers with procedures such as periodic tests and checks of equipment, preparations for arrival and departure, changeover of steering modes, signalling, communications, alarm system, manoeuvring, emergencies and logbook entries.

## 14    Cargo operation

**14.1**    The PSCO may determine if ship's personnel assigned with specific duties related to the cargo and cargo equipment are familiar with those duties, any dangers posed by the cargo and with the measures to be taken in such a context. This will require the availability of all relevant cargo information as required by SOLAS regulation VI/2.

**14.2**    With respect to the carriage of solid bulk cargoes, the PSCO should verify, as appropriate, that cargo loading is performed in accordance with a ship's loading plan and unloading in accordance with a ship's unloading plan agreed by the ship and the terminal, taking into account the information provided by the loading instrument, where fitted.

**14.3**    The PSCO, when appropriate, may determine whether the responsible crew members are familiar with the relevant provisions of the International Maritime Solid Bulk Cargoes Code (IMSBC Code), particularly those concerning moisture limits and trimming of the cargo, the Code of Safe Practice for Ships Carrying Timber Deck Cargoes (TDC Code 2011) and the Code of Safe Practice for Cargo Stowage and Securing.

**14.4**    Some solid materials transported in bulk can present a hazard during transport because of their chemical nature or physical properties. Section 2 of the IMSBC Code gives general precautions. Section 4 of the IMSBC Code contains the obligation imposed on the shipper to provide all necessary information to ensure a safe transport of the cargo. The PSCO may determine whether all relevant details, including all relevant certificates of tests, have been provided to the master by the shipper.

**14.5**    For some cargoes, such as cargoes which are subject to liquefaction, special precautions are given (see section 7 of the IMSBC Code). The PSCO may determine whether all precautions are met with special attention for the stability of those ships engaged in the transport of cargoes subject to liquefaction and solid hazardous waste in bulk.

**14.6**    Officers responsible for cargo handling and operation and key crew members of oil tankers, chemical tankers and liquefied gas carriers should be familiar with the cargo and cargo equipment and with the safety measures as stipulated in the relevant sections of the IBC and IGC Codes.

**14.7**    For the carriage of grain in bulk, reference is made to part C of chapter VI of SOLAS and the International Code for the Safe Carriage of Grain in Bulk (Grain Code).

**14.8**    The PSCO may determine whether the operations and loading manuals include all the relevant information for safe loading and unloading operations in port as well as in transit conditions.

## 15    Operation of the machinery

**15.1**    The PSCO may determine if responsible ship's personnel are familiar with their duties related to operating essential machinery, such as:

 **.1**    emergency and standby sources of electrical power;

 **.2**    auxiliary steering gear;

 **.3**    bilge and fire pumps; and

 **.4**    any other equipment essential in emergency situations.

**15.2**    The PSCO may verify whether the responsible ship's personnel are familiar with, inter alia:

 **.1**    emergency generator:

  **.1**    actions which are necessary before the engine can be started;

  **.2**    different possibilities to start the engine in combination with the source of starting energy; and

  **.3**    procedures when the first attempts to start the engine fail; and

 **.2**    standby generator engine:

  **.1**    possibilities to start the standby engine, automatic or by hand;

  **.2**    blackout procedures; and

  **.3**    load-sharing system.

**15.3**    The PSCO may verify whether the responsible ship's personnel are familiar with, inter alia:

 **.1**    which type of auxiliary steering gear system applies to the ship;

 **.2**    how it is indicated which steering gear unit is in operation; and

 **.3**    what action is needed to bring the auxiliary steering gear into operation.

**15.4**    The PSCO may verify whether the responsible ship's personnel are familiar with, inter alia:

  **.1**    bilge pumps:

  **.1**    number and location of bilge pumps installed on board the ship (including emergency bilge pumps);

  **.2**    starting procedures for all these bilge pumps;

  **.3**    appropriate valves to operate; and

  **.4**    most likely causes of failure of bilge pump operation and their possible remedies; and

  **.2**    fire pumps:

  **.1**    number and location of fire pumps installed on board the ship (including the emergency fire pump);

  **.2**    starting procedures for all these pumps; and

  **.3**    appropriate valves to operate.

**15.5**    The PSCO may verify whether the responsible ship's personnel are familiar with, inter alia:

  **.1**    starting and maintenance of lifeboat engine and/or rescue boat engine;

  **.2**    local control procedures for those systems which are normally controlled from the navigating bridge;

  **.3**    use of the emergency and fully independent sources of electrical power of radio installations;

  **.4**    maintenance procedures for batteries;

  **.5**    emergency stops, fire detection system and alarm system operation of watertight and fire doors (stored energy systems); and

  **.6**    change of control from automatic to manual for cooling water and lube oil systems for main and auxiliary engines.

## 16    Manuals, instructions, etc.

**16.1**    The PSCO may determine if the appropriate crew members are able to understand the information given in manuals, instructions, etc., relevant to the safe condition and operation of the ship and its equipment and that they are aware of the requirements for maintenance, periodic testing, training, drills and recording of logbook entries.

**16.2**    The following information, inter alia, should be provided on board and PSCOs may determine whether it is in a language or languages understood by the crew and whether crew members concerned are aware of the contents and are able to respond accordingly:

  **.1**    instructions concerning the maintenance and operation of all the equipment and installations on board for the fighting and containment of fire should be kept under one cover, readily available in an accessible position;

  **.2**    clear instructions to be followed in the event of an emergency should be provided for every person on board;

  **.3**    illustrations and instructions in appropriate languages should be posted in passenger cabins and be conspicuously displayed at muster stations and other passenger spaces to inform passengers of their muster station, the essential action they must take in an emergency and the method of donning lifejackets;

  **.4**    posters and signs should be provided on or in the vicinity of survival craft and their launching controls and shall illustrate the purpose of controls and the procedures for operating the appliance and give relevant instructions or warnings;

  **.5**    instructions for on-board maintenance of life-saving appliances;

.6    training manuals should be provided in each crew mess room and recreation room or in each crew cabin; the training manual, which may comprise several volumes, should contain instructions and information, in easily understood terms illustrated wherever possible, on the life-saving appliances provided in the ship and on the best method of survival;

.7    Shipboard Oil Pollution Emergency Plan for noxious liquid substances in accordance with regulation 37 of MARPOL Annex I, or Shipboard Marine Pollution Emergency Plan for noxious liquid substances in accordance with regulation 17 of MARPOL Annex II, where applicable; and

.8    stability booklet, associated stability plans and stability information.

## 17    Oil and oily mixtures from machinery spaces

**17.1**    The PSCO may determine if all operational requirements of Annex I of MARPOL have been met, taking into account:

.1    the quantity of oil residues generated;

.2    the capacity of sludge and bilge water holding tank; and

.3    the capacity of the oily water separator.

**17.2**    An inspection of the Oil Record Book should be made. The PSCO may determine if reception facilities have been used and note any alleged inadequacy of such facilities.

**17.3**    The PSCO may determine whether the responsible officer is familiar with the handling of sludge and bilge water. The relevant items from the guidelines for systems for handling oily wastes in machinery spaces of ships may be used as guidance. Taking into account the above, the PSCO may determine if the ullage of the sludge tank is sufficient for the expected generated sludge during the next intended voyage. The PSCO may verify that, in respect of ships for which the Administration has waived the requirements of regulations 14(1) and (2) of MARPOL Annex I, all oily bilge water is retained on board for subsequent discharge to a reception facility.

**17.4**    When reception facilities in other ports have not been used because of inadequacy, the PSCO should advise the master to report the inadequacy of the reception facility to the ship's flag State, in conformity with the Format for reporting alleged inadequacies of port reception facilities (MEPC.1/Circ.834, appendix 1 of the annex), as may be amended.

## 18    Loading, unloading and cleaning procedures for cargo spaces of tankers

**18.1**    The PSCO may determine if all operational requirements of Annexes I or II of MARPOL have been met taking into account the type of tanker and the type of cargo carried, including the inspection of the Oil Record Book and/or Cargo Record Book. The PSCO may determine if the reception facilities have been used and note any alleged inadequacy of such facilities.

**18.2**    For the control on loading, unloading and cleaning procedures for tankers carrying oil, reference is made to paragraphs 3.1 to 3.4 in appendix 5 where guidance is given for the inspection of crude oil washing (COW) operations. In appendix 3, the PSCO may find detailed guidelines for in-port inspection of crude oil washing procedures.

**18.3**    For the control on loading, unloading and cleaning procedures for tankers carrying noxious liquid substances, reference is made to paragraphs 4.1 to 4.9 in appendix 5 where guidance is given for the inspection of unloading, stripping and prewash operations. In appendix 4 more detailed guidelines for these inspections are given.

**18.4**    When reception facilities in other ports have not been used because of inadequacy, the PSCO should advise the master to report the inadequacy of the reception facility to the ship's flag State, in conformity with MEPC/Circ.834 (April 2014).

**18.5**    When a ship is permitted to proceed to the next port with residues of noxious liquid substances on board in excess of those permitted to be discharged into the sea during the ship's passage, it should be ascertained that the residues can be received by that port. At the same time that port should be informed if practicable.

# 19    Dangerous goods and harmful substances in packaged form

**19.1**    The PSCO may determine if the required shipping documents for the carriage of dangerous goods and harmful substances carried in packaged form are provided on board and whether the dangerous goods and harmful substances are properly stowed and segregated, and the crew members are familiar with the essential action to be taken in an emergency involving such packaged cargo (see SOLAS regulation VII/3).

**19.2**    Ship types and cargo spaces of ships of over 500 gross tonnage built on or after 1 September 1984 and ship types and cargo spaces of ships of less than 500 gross tonnage built on or after 1 February 1992 are to fully comply with the requirements of SOLAS chapter II-2. Administrations may reduce the requirements for ships of less than 500 gross tonnage but such reductions shall be recorded in the document of compliance. A document of compliance is not required for ships which only carry class 6.2, class 7 or dangerous goods in limited quantities.

**19.3**    Annex III of MARPOL contains requirements for the carriage of harmful substances in packaged form which are identified in the IMDG Code as marine pollutants. Cargoes which are determined to be marine pollutants should be labelled and stowed in accordance with Annex III of MARPOL.

**19.4**    The PSCO may determine whether a Document of Compliance is on board and whether the ship's personnel are familiar with this document provided by the Administration as evidence of compliance of construction and equipment with the requirements. Additional control may consist of:

 .1    whether the dangerous goods have been stowed on board in conformity with the Document of Compliance, using the dangerous goods manifest or the stowage plan, required by SOLAS chapter VII; this manifest or stowage plan may be combined with the one required under Annex III of MARPOL;

 .2    whether inadvertent pumping of leaking flammable or toxic liquids is not possible in case these substances are carried in under-deck cargo spaces; or

 .3    determining whether the ship's personnel are familiar with the relevant provisions of the Medical First Aid Guide and Emergency Procedures for Ships Carrying Dangerous Goods.

# 20    Garbage

**20.1**    The PSCO may determine if all operational requirements of Annex V of MARPOL have been met. The PSCO may determine if the reception facilities have been used and note any alleged inadequacy of such facilities.

**20.2**    Guidelines for the implementation of Annex V of MARPOL were approved by MEPC at its twenty-ninth session and have been amended on numerous occasions. The Guidelines can be found within the consolidated text of MARPOL Annex V. One of the objectives of these Guidelines is to assist ship operators complying with the requirements set forth in Annex V and domestic laws.

**20.3**    The PSCO may determine whether:

 .1    ship's personnel are aware of these Guidelines, in particular section 2, Garbage management, and section 4, Shipboard garbage handling and storage procedures; and

 .2    ship's personnel are familiar with the disposal and discharge requirements under Annex V of MARPOL inside and outside a special area and are aware of the areas determined as special areas under Annex V of MARPOL.

**20.4**    When reception facilities in other ports have not been used because of inadequacy, the PSCO should advise the master to report the inadequacy of the reception facility to the ship's flag State, in conformity with MEPC/Circ.834, as may be amended.

## 21    Sewage

**21.1**    The PSCO may determine:

.1    if all operational requirements of Annex IV of MARPOL have been met; the PSCO may determine if the sewage treatment system, comminuting and disinfecting system or holding tank has been used and note any alleged inadequacy of the system or holding tank; and

.2    that appropriate ship's personnel are familiar with the correct operation of the sewage treatment system, comminuting and disinfecting system or holding tank.

**21.2**    The PSCO may determine whether appropriate ship's personnel are familiar with the discharge requirements of regulation 11 of MARPOL Annex IV.

**21.3**    When reception facilities in other ports have not been used because of inadequacy, the PSCO should advise the master to report the inadequacy of the reception facility to the ship's flag State, in conformity with the waste reception facility reporting requirements (MEPC.1/Circ.834, as may be amended).

## 22    Air pollution prevention

The PSCO may determine whether:

.1    the master or crew is familiar with the procedures to prevent emissions of ozone-depleting substances;

.2    the master or crew is familiar with the proper operation and maintenance of diesel engines, in accordance with their Technical Files;

.3    the master or crew has undertaken the necessary fuel changeover procedures or equivalent, associated with demonstrating compliance within a SOX emission control area;

.4    the master or crew is familiar with the garbage screening procedure to ensure that prohibited garbage is not incinerated;

.5    the master or crew is familiar with the operation of the shipboard incinerator, as required by regulation 16.2 of MARPOL Annex VI, within the limit provided in appendix IV to the Annex, in accordance with the operational manual;

.6    the master or crew recognizes the regulation of emissions of volatile organic compounds (VOCs), when the ship is in ports or terminals under the jurisdiction of a Party to the 1997 Protocol to MARPOL in which VOCs emissions are to be regulated, and is familiar with the proper operation of a vapour collection system approved by the Administration (in case the ship is a tanker as defined in regulation 2.27 of MARPOL Annex VI); and

.7    the master or crew is familiar with bunker delivery procedures in respect of bunker delivery notes and retained samples as required by regulation 18 of MARPOL Annex VI.

# Appendix 8
*Guidelines for port State control officers on the ISM Code*

## 1 General

**1.1** The International Safety Management Code (ISM Code) was adopted by the Assembly at its eighteenth session by resolution A.741(18) and was amended by resolutions MSC.104(73), MSC.179(79), MSC.195(80), MSC.273(85) and MSC.353(92). The ISM Code has been made mandatory through SOLAS regulation IX/3.

**1.2** The Administration is responsible for verifying compliance with the requirements of the ISM Code and issuing Documents of Compliance to companies and Safety Management Certificates to ships. This verification is carried out by the Administration or a recognized organization (RO).

**1.3** Port State control officers (PSCOs) do not perform safety management audits. PSCOs conduct inspections of ship, which are a sampling process and give a snapshot of the vessel on a particular day.

## 2 Goals and purpose

**2.1** The guidelines provide guidance to PSCOs for the harmonized application of related technical or operational deficiencies found in relation to the ISM Code during a PSC inspection.

## 3 Application

**3.1** The ISM Code applies to the following types of ships engaged in international voyages:

.1 all passenger ships including passenger high-speed craft;

.2 oil tankers, chemical tankers, gas carriers, bulk carriers and cargo high-speed craft of 500 gross tonnage and above; and

.3 other cargo ships and self-propelled mobile offshore drilling units (MODUs) of 500 gross tonnage and above.

**3.2** For establishing the applicability of SOLAS chapter IX and the ISM Code, "gross tonnage" means the gross tonnage of the ship as determined under the provisions of the International Convention on the Tonnage Measurement of Ships, 1969, and as stated on the International Tonnage Certificate of the ship.

**3.3** The ISM Code does not apply to government-operated ships used for non-commercial purposes.

## 4 Relevant documentation

**4.1** Applicable documentation for these guidelines is as follows:

.1 SOLAS;

.2 ISM Code;

.3 Copy of the Interim DOC, or copy of the DOC;

.4 Interim SMC, or SMC; and

.5 MSC/Circ.1059-MEPC/Circ.401, as may be amended.

# 5 Definitions and abbreviations

*SOLAS*  International Convention for the Safety of Life at Sea, 1974, as amended.

*ISM Code*  International Safety Management Code: The International Management Code for the Safe Operation of Ships and for Pollution Prevention adopted by the Organization by resolution A.741(18), as amended.

*Procedures for port State control*  Procedures for port State control, 2017, as adopted by resolution A.1119(30), as may be amended.

*Company*  The owner of the ship or any other organization or person such as the manager, or the bareboat charterer, who has assumed the responsibility for operation of the ship from the shipowner and who, on assuming such responsibility, has agreed to take over all duties and responsibility imposed by the Code.

*Administration*  The Government of the State whose flag the ship is entitled to fly.

*DOC*  Document of Compliance: A document issued to a company which complies with the requirements of the ISM Code.

*SMC*  Safety Management Certificate: A document issued to a ship which signifies that the company and its shipboard management operate in accordance with the approved safety management system.

*SMS*  Safety Management System: A structured and documented system enabling company personnel to implement effectively the company safety and environmental protection policy.

*Objective evidence*  Quantitative or qualitative information, records or statements of fact pertaining to safety or to the existence and implementation of a safety management system element, which is based on observation, measurement or test and which can be verified.

*Valid certificate*  A certificate that has been issued directly by a Party to a relevant convention or on its behalf by a recognized organization and contains: accurate and effective dates; meets the provisions of the relevant convention; and, with which the particulars of the ship, her crew and her equipment correspond.

*PSC*  Port State control.

*PSCO*  Port State control officer.

*RO*  Recognized organization: An organization recognized by the Administration.

*MODU*  Mobile offshore drilling unit.

# 6 Inspection of ship

## 6.1 Initial inspection

**6.1.1** Initial inspection should be carried out in accordance with the Procedures for port State control.

**6.1.2** During the initial PSC inspection, the PSCO should verify that the ship carries the ISM certificates according to SOLAS chapter IX and the ISM Code by examining the copy of the DOC and the SMC, for which the following points are to be considered:

.1 A copy of the DOC should be on board. However, according to SOLAS, the copy of the DOC is not required to be authenticated or certified. The copy of the DOC should have the required endorsements.

.2 The SMC is not valid unless the operating company holds a valid DOC for that ship type. The ship type in the SMC should be included in the DOC and the company's particulars should be the same on both the DOC and the SMC. The SMC should have the required endorsements.

.3 The validity of an Interim DOC should not exceed a period of 12 months. The validity of an Interim SMC should not exceed a period of six months. In special cases, the Administration, or at the request of the Administration another Government, may extend the validity of the Interim SMC for a period which should not exceed six months from the date of expiry.

.4 ROs may issue a short-term DOC or SMC not exceeding five months, while the full term certificate is being prepared in accordance with their internal procedures. If a renewal verification has been completed and a new SMC cannot be issued or placed on board the ship before the expiry date of the existing certificate, the Administration or RO may endorse the existing certificate. Such a certificate should be accepted as valid for a further period which should not exceed five months from the expiry date.

.5 If a ship at the time when an SMC expires is not in a port in which it is to be verified, the Administration may extend the period of validity of the SMC but this extension should be granted only for the purpose of allowing the ship to complete her voyage to the port in which it is to be verified, and then only in cases where it appears proper and reasonable to do so.

.6 No SMC should be extended for a period of longer than three months, and the ship to when an extension is granted should not, on her arrival in the port in which it is to be verified, be entitled by virtue of such extension to leave that port without having a new SMC. When the renewal verification is completed, the new SMC should be valid to a date not exceeding five years from the expiry date of the existing SMC before the extension was granted.

.7 If no technical or operational related deficiencies are found during an initial inspection carried out in accordance with the Procedures for port State control and guidelines, there is no need to consider the ISM aspect.

## 6.2 Clear grounds

**6.2.1** Since the PSCO is not carrying out a safety management audit of the SMS during a PSC inspection, the term "clear grounds" is not applicable in this context.

**6.2.2** Clear grounds and the subsequent more detailed inspection only exist for technical or operational-related deficiencies.

## 6.3 More detailed inspection

**6.3.1** If a more detailed inspection for technical or operational-related deficiencies is carried out, this should be done in accordance with the Procedures for port State control. Any technical and/or operational-related deficiencies found during this inspection should be, individually or collectively considered by the PSCO, using their professional judgement, to indicate that either:

.1 these do not show a failure, or lack of effectiveness, of the implementation of the ISM Code; or

    **.2**    there is a failure, or lack of effectiveness, of the implementation of the ISM Code; or

    **.3**    there is a serious failure, or lack of effectiveness, of the implementation of the ISM Code.

**6.3.2**    If an outstanding ISM related deficiency from a previous PSC inspection exists and the current PSC inspection is more than three months later:

    **.1**    the PSCO will verify that an internal safety audit has been performed; the content of the internal safety audit report should not be evaluated; and

    **.2**    having reference to the previous PSC inspection report, the PSCO will examine the technical and/or operational areas in which deficiencies designated with "ISM" are noted.

# 7    Follow-up action

## 7.1    Technical, operational and ISM-related deficiencies

**7.1.1**    The principles outlined in the Procedures for port State control with regard to reporting and rectification of technical or operational-related deficiencies, and detention and release of the ship are applicable.

**7.1.2**    If there are technical or operational-related deficiencies reported which:

    **.1**    do not show a failure, or lack of effectiveness, of the implementation of the ISM Code – no ISM-related deficiency should be reported in the PSC inspection report;

    **.2**    individually or collectively do not warrant the detention of the ship but indicate a failure, or lack of effectiveness, of the implementation of the ISM Code – report an ISM-related deficiency in the PSC inspection report with the requirement of an internal safety audit and corrective action within three months; and

    **.3**    individually or collectively lead to detention of the ship and indicate a serious failure, or lack of effectiveness, of the implementation of the ISM Code – report an ISM-related deficiency in the PSC inspection report with the requirement that a safety management audit has to be carried out by the Administration or the RO before the ship may be released from her detention.

    **Note:** Where the PSCO considers one or more technical and/or operational deficiency(s) is ISM-related this should be recorded as only one ISM deficiency.

**7.1.3**    If an outstanding ISM-related deficiency (to be rectified within three months) from a previous PSC inspection exists and no objective evidence can be provided by the master of the ship, during the current PSC inspection more than three months later, that an internal safety audit has been performed, any further action will be taken based on the professional judgement of the PSCO and may warrant the detention of the ship.

## 7.2    Deficiencies not warranting detention

**7.2.1**    Minor typing errors in the DOC or the SMC should be reported in the PSC inspection report as a technical deficiency with the certificates and not an ISM-related deficiency.

**7.2.2**    If technical and/or operational-related deficiencies are found and reported during the PSC inspection which do not warrant detention but in the professional judgement of the PSCO provide objective evidence of a failure, or lack of effectiveness, of the implementation of the ISM Code, this should be reported additionally in the PSC inspection report as an ISM-related deficiency.

## 7.3    Deficiencies warranting detention

The following are deficiencies which may warrant detention:

    **.1**    deficiencies of a technical and/or operational nature which individually or collectively provide objective evidence of a serious failure, or lack of effectiveness, of the implementation of the ISM Code;

    **.2**    there is no SMC, interim SMC and/or copy of the DOC or interim DOC on board the ship;

**.3** there is no valid SMC or interim SMC on board;

**.4** the SMC intermediate verification is overdue;

**.5** the SMC has expired and there is no objective evidence of an extension issued by the Administration; or where the SMC has been withdrawn by the Administration;

**.6** the DOC or interim DOC has expired or been withdrawn;

**.7** the ship type as indicated on the SMC or interim SMC is not listed on the DOC or interim DOC;

**.8** evidence of the DOC annual verification is not available on board;

**.9** the certificate number on the copy of the DOC and the endorsement pages are not the same; and

**.10** the company name, the company address or the issuing Government authority on the DOC or interim DOC is not the same as on the SMC or interim SMC.

# 8 Reporting

## 8.1 Technical and operational-related deficiencies

**8.1.1** All technical and/or operational-related deficiencies should be recorded as an individual deficiency in the PSC inspection report according to the Procedures for port State control.

**8.1.2** Technical-related deficiency with the defective item DOC/SMC or interim DOC/SMC should be recorded in the PSC inspection report as a certificate deficiency.

## 8.2 ISM-related deficiency

Where the PSCO has considered the technical and/or operational-related deficiencies found and concluded these provide objective evidence of a (serious) failure, or lack of effectiveness of the implementation of the ISM Code, an ISM-related deficiency should be recorded in the PSC inspection report.

# Appendix 9
*Guidelines for port State control related to LRIT*

## 1    Purpose

These Guidelines are intended to provide basic guidance to PSCOs to verify compliance with the requirements of SOLAS for Long Range Identification and Tracking (LRIT).

## 2    Application

**2.1**    LRIT equipment is required by SOLAS regulation V/19-1, and resolution MSC.263(84) requires all passenger ships, cargo ships (including high-speed craft) over 300 gross tonnage and mobile offshore drilling units (MODUs) to send LRIT position information at least every 6 hours. Ships fitted with an automatic identification system (AIS) and operated exclusively within sea area A1 are not required to comply with LRIT. Sea area A1 is defined by SOLAS regulation IV/2.1.12 as "an area within the radiotelephone coverage of at least one VHF coast station in which continuous DSC alerting is available, as may be defined by a Contracting Government".

**2.2**    SOLAS Contracting Governments are expected to maintain an LRIT data centre, either on a national basis, or on a regional or cooperative basis with other flag States, and notify the IMO of it. In turn, these LRIT data centres will forward, upon request, LRIT information from ships entitled to fly their flag, to other SOLAS Contracting Governments through the International LRIT Data Exchange. Port States are entitled to request the LRIT information from foreign ships that have indicated their intention to enter a port, port facility or place under its jurisdiction.

**2.3**    In most cases a stand-alone Inmarsat C or Inmarsat mini-C terminal used for GMDSS or ship security alert system will function as the LRIT terminal, but other equipment may be employed for the LRIT function (example Inmarsat D+ or Iridium).

## 3    Inspection of ships required to carry LRIT equipment

### 3.1    Initial inspection

**3.1.1**    The PSCO should first establish the sea area the ship is certified to operate in. This verification should ensure that the ship is subject to the LRIT regulation in relation to its ship type and tonnage. After the certificate check, the PSCO should verify that:

  .1    the Record of Equipment (Form E, P or C) indicates LRIT as required, if applicable;[*]

  .2    a Statement of Conformity/Conformance Test Report (see MSC.1/Circ.1307) is on board; and

  .3    the equipment identified by the ship's representative as the designated LRIT terminal is switched on.[†]

**3.1.2**    In case of recent transfer of flag, the PSCO may further ensure that:

  .1    a conformance test report has been re-issued if the new flag State does not recognize the issuing body of the existing conformance test report; or

  .2    a new conformance test has been carried out by the application service provider (ASP) on behalf of the Administration before issuance of a new test report and certificate.

---

[*] Noting that a Record of Equipment is required for cargo ships greater than 500 gross tonnage and passenger ships.

[†] **Note:** In exceptional circumstances and for the shortest duration possible LRIT is capable of being switched off or may transmit less frequently (SOLAS regulation V/19-1.7.2 and resolution MSC.263(84), paragraph 4.4.1).

## 3.2    Clear grounds

Conditions which may warrant a more detailed inspection of equipment used for LRIT may comprise the following:

.1    defective main or emergency source of energy;

.2    information or indication that LRIT equipment is not functioning properly;

.3    ship does not hold conformance test report; and

.4    the "record of navigational activities" indicates that the LRIT installation has been switched off and that this has not been reported to the flag Administration as required by SOLAS regulation V/19-1.7.2.

## 3.3    More detailed inspection

**3.3.1**    In case of doubt or reports of malfunctioning of the LRIT installation, the flag Administration may be contacted to determine if the ship's LRIT information has been reliably relayed to the LRIT data centre.

**3.3.2**    If any issues are identified at the initial inspection, a more detailed inspection of equipment used for LRIT may comprise the following:

.1    verification of the power supply, which should be connected to the main source of energy and the emergency source of energy – there is no requirement for an uninterrupted power source; if the LRIT is part of the GMDSS radio-installation, the power supply should conform to GMDSS regulations;

.2    inspection of the "record of navigational activities" log to establish if and when the installation has been switched off and if this has been reported to the flag Administration (SOLAS regulation V/19-1.7.2 and resolution MSC.263(84), paragraph 4.4.1); and

.3    ensuring that any conformance test report is issued on behalf of the flag State, even by itself or by an authorized application service provider (see MSC.1/Circ.1377/Rev.11 and further version as shown in GISIS), available for a ship that has an LRIT installation.

# 4    Deficiencies warranting detention

**4.1**    A PSCO should use professional judgement to determine whether to detain the ship until any noted deficiencies are corrected or to permit a vessel to sail with deficiencies.[*]

**4.2**    In order to assist the PSCO in the use of these Guidelines, the following deficiencies should be considered to be of such nature that they may warrant the detention of a ship:

.1    absence of a valid LRIT Conformance test report; and

.2    the master or the responsible officer is not familiar with essential shipboard operational procedures relating to LRIT.

**4.3**    Taking into account the guidance found in the Guidance on the implementation of the LRIT system (MSC.1/Circ.1298), PSCOs are also advised that ships should not be detained if the LRIT installation on board works, but the shore-side installation or organization is not able to receive, relay or process the information.

**4.4**    PSCOs are advised that a flag State may issue a short-term certificate; this could happen if, following a successful inspection for the issuance of a Conformity Test report, the ASP has not been able to issue a document yet, or if the ASP is not able to perform a conformance test in due time upon the request of the shipowner.

---

[*] SOLAS regulation V/16: "Whilst all reasonable steps shall be taken to maintain the equipment required by this chapter in efficient working order, malfunctions of that equipment shall not be considered as making the ship unseaworthy or as a reason for delaying the ship in ports where repair facilities are not readily available, provided suitable arrangements are made by the master to take the inoperative equipment or unavailable information into account in planning and executing a safe voyage to a port where repairs can take place."

# Appendix 10
## *Guidelines for port State control under the 1969 Tonnage Convention*

**1**    The International Convention on Tonnage Measurement of Ships, 1969, which came into force on 18 July 1982, applies to:

.1    new ships, i.e. ships the keels of which were laid on or after 18 July 1982; and

.2    existing ships, i.e. ships the keels of which were laid before 18 July 1982, as from 18 July 1994,

except that for the purpose of application of the SOLAS, MARPOL and STCW Conventions, the following interim schemes indicated in paragraph 2 may apply.

**2**    In accordance with the interim schemes adopted by the Organization,[*] the Administration may, at the request of the shipowner, use the gross tonnage determined in accordance with national rules prior to the coming into force of the 1969 Tonnage Convention, for the following ships:

.1    for the purpose of SOLAS:

.1    ships the keels of which were laid before 1 January 1986;

.2    in respect of SOLAS regulation IV/3, ships the keels of which were laid on or after 1 January 1986 but before 18 July 1994; and

.3    cargo ships of less than 1,600 tons gross tonnage (as determined under the national tonnage rules) the keels of which were laid on or after 1 January 1986 but before 18 July 1994; and

.2    for the purpose of MARPOL, ships of less than 400 tons gross tonnage (as determined under the national tonnage rules) the keels of which were laid before 18 July 1994.

**3**    For ships to which the above interim schemes apply, a statement to the effect that the gross tonnage has been measured in accordance with the national tonnage rules should be included in the "REMARKS" column of the International Tonnage Certificate (1969) and in the footnote to the figure of the gross tonnage in the relevant SOLAS and MARPOL certificates.

**4**    The PSCO should take the following actions as appropriate when deficiencies are found in relation to the 1969 Tonnage Convention:

.1    if a ship does not hold a valid International Tonnage Certificate (1969), the ship loses all privileges of the 1969 Tonnage Convention, and the flag State should be informed without delay;

.2    if the required remarks and footnote are not included in the relevant certificates on ships to which the interim schemes apply, this deficiency should be notified to the master; and

.3    if the main characteristics of the ship differ from those entered on the International Tonnage Certificate (1969), so as to lead to an increase in the gross tonnage or net tonnage, the flag State should be informed without delay.

**5**    The control provisions of article 12 of the 1969 Tonnage Convention do not include the provision for detention of a ship holding a valid International Tonnage Certificate (1969).

---

[*] Resolutions A.494(XII) in respect of SOLAS, A.540(13) in respect of STCW 78, and A.541(13) in respect of MARPOL.

# Appendix 11
*Guidelines for port State control officers on certification of seafarers, manning and hours of rest*

## 1    General

The International Convention for the Safety of Life at Sea (SOLAS) was adopted in 1974 and entered into force in 1980. Similarly, the International Convention on Standards of Training, Certification and Watchkeeping for Seafarers (STCW) was adopted in 1978 and entered into force in 1984. Both have been amended several times since their entry into force.

## 2    Goals and purpose

These guidelines are intended to provide guidance for a harmonized approach of port State control (PSC) inspections in compliance with SOLAS regulations V/14 (manning) and I/2 (seafarer certification) and chapter VIII (hours of rest) of the STCW Convention, as amended.

## 3    Application

**3.1**    SOLAS regulation V/14.2 only applies to ships covered by chapter I of SOLAS. The STCW Convention as amended applies to seafarers serving on board seagoing ships. The STCW Code is divided into a mandatory part A and a non-mandatory part B. Part B of the STCW Code is not applicable during the inspection.

**3.2**    All passenger ships regardless of size and all other ships of 500 gross tonnage or more should have a minimum safe manning document or equivalent on board issued by the flag State.

**3.3**    Any new or single deficiency which is either a deficiency related to SOLAS, STCW or other IMO conventions, should preferably be registered with these conventions references.

## 4    Relevant documentation

The documentation required for the inspection referred to in these Guidelines consists of:

*Seafarer certification*

    **.1**    certificate of competency;

    **.2**    certificate of proficiency;

    **.3**    endorsement attesting the recognition of a certificate (flag State endorsement);

    **.4**    documentary evidence (passenger ships only);

    **.5**    medical certificate;

*Manning*

    **.6**    minimum safe manning document;

    **.7**    muster list;

*Hours of rest*

    **.8**    table of ship working arrangements and/or watch schedule; and

    **.9**    records of daily hours of rest.

## 5    Definitions and abbreviations

**5.1**    Certificate of Competency means a certificate issued and endorsed for masters, officers and Global Maritime Distress and Safety System (GMDSS) radio operators in accordance with the provisions of chapters II, III, IV or VII of the STCW Convention and entitling the lawful holder thereof to serve in the capacity and perform the functions involved at the level of responsibility specified therein.

**5.2**    Certificate of Proficiency means a certificate, other than a certificate of competency issued to a seafarer, stating that the relevant requirements of training, competencies or seagoing service in the STCW Convention have been met.

**5.3**    Documentary evidence means documentation, other than a Certificate of Competency or Certificate of Proficiency, used to establish that the relevant requirements of the STCW Convention, 1978, as amended, have been met. The only documentary evidence required under the STCW Convention, 1978, as amended, is issued to personnel meeting the mandatory minimum requirements for the training and qualifications of masters, officers, ratings and other personnel on passenger ships (regulation V/2).

**5.4**    The following abbreviations have been used:

   **.1**    CoC (Certificate of Competency);

   **.2**    CoP (Certificate of Proficiency); and

   **.3**    MSMD (minimum safe manning document).

## 6    Inspection of ship

### 6.1    Pre-boarding preparation

**6.1.1**    Taking into account the type, size, engine power and other particulars of the ship, the port State control officer (PSCO) should be aware of the relevant requirements of SOLAS regulation V/14 and the STCW Convention.

**6.1.2**    The PSCO should be aware that resolutions are non-mandatory documents and not applicable during a PSC inspection.

**6.1.3**    The PSCO should also identify if the flag State is a Party to the STCW Convention, as amended. If the flag State is not a Party to the Convention or is a Party but not listed in MSC.1/Circ.1163, as amended, a more detailed inspection should be carried out.

### 6.2    Initial inspection

*Seafarer certificates and documents*

**6.2.1**    The PSCO should examine the applicable documents, found in section 4.

**6.2.2**    The inspection should be limited to verification that seafarers serving on board, who are required to be certificated, hold the appropriate Certificates of Competency, Certificates of Proficiency and documentary evidence issued in accordance with chapters II, III, IV, V, VI and VII of the STCW Convention, 1978, as amended, as well as their relevant flag State endorsement, valid dispensation, or documentary proof that an application for an endorsement has been submitted to the flag State Administration, where applicable. These documents are evidence of having successfully completed all required training and that the required standard of competence has been achieved.

**6.2.3**    During the verification of the seafarers' certificates and documents, the PSCO should confirm that they are applicable to the ship's characteristics, operation and their position on board.

**6.2.4**    In accordance with the provision of article VI paragraph 2 of the STCW Convention, certificates for masters and officers should be endorsed by the issuing Administration in the form prescribed in regulation I/2 of the annex to the convention.

**6.2.5**  The certificates may be issued as one certificate with the required endorsement incorporated. If so incorporated, the form used should be that set forth in section A-I/2, paragraph 1 of the STCW Code.

**6.2.6**  The endorsement may also be issued as a separate document. If so, the form used should be that set forth in section A-I/2, paragraph 2 of the STCW Code.

**6.2.7**  However, Administrations may use a format for certificates and endorsements different from those given in section A-I/2 of the STCW Code, provided that, at a minimum, the required information is provided in Roman characters and Arabic figures. Permitted variations to the format are set out in section A-I/2, paragraph 4 of the STCW Code.

**6.2.8**  Certificates and endorsements issued as separate documents should each be assigned a unique number, except that endorsements attesting the issuance of a certificate may be assigned the same number as the certificate concerned, provided that number is unique.

**6.2.9**  Certificates and endorsements issued as separate documents should include a date of expiry. The date of expiry on an endorsement issued as a separate document should not exceed 5 years from the date of issue and may never exceed the date of expiry on the certificate.

**6.2.10**  A CoP issued to a master or an officer in accordance with regulation V/1-1 or V/1-2, as well as a CoC that has been issued by a State other than the flag State of the ship in which the seafarer is engaged, is required to be recognized by the ship's flag State. If the PSCO identifies that the flag State has recognized a CoC or CoP from a Party not listed in MSC.1/Circ.1163, as amended, clarification should be sought from the flag Administration. According to regulation I/10, paragraph 4 of the STCW Convention, certificates issued by or under the authority of a non-Party shall not be recognized by the ship's flag State Administration.

**6.2.11**  An Administration which recognizes under regulation I/10 a CoC or CoP issued to masters and officers should endorse that certificate to attest to its recognition. The form of the endorsement should be that found in section A-I/2 paragraph 3 of the STCW code.

**6.2.12**  Incorrect wording or missing information may be a cause for suspicion regarding fraudulent certificates or endorsements.

**6.2.13**  Endorsements attesting to the recognition of a certificate should each be assigned a unique number, however they may be assigned the same number as the certificate concerned, provided that number is unique.

**6.2.14**  Endorsements attesting to the recognition of a certificate should include a date of expiry. The date of expiry on an endorsement attesting to the recognition may never exceed the date of expiry on the certificate being recognized.

**6.2.15**  The capacity in which the holder of a certificate is authorized to serve should be identified in the form of endorsement in terms identical to those used in the applicable safe manning requirements of the Administration. This may result in slight variations of terminology between the original CoC and the endorsement to the recognition.

**6.2.16**  Seafarers must have their original CoC on board as well as any original endorsements to the recognition. An endorsement attesting the recognition of a certificate should not entitle a seafarer to serve in a higher capacity than the original CoC.

**6.2.17**  If circumstances require it, a flag State Administration may permit a seafarer to serve for a period not exceeding three months on ships entitled to fly its flag while holding a valid CoC issued by another party and valid for service on that party's ships. If such a situation exists, documentary proof must be readily available that an application for endorsement has been made to the Administration of the flag State. This is often referred to as the confirmation of receipt of application (CRA). This provision allows Administrations to permit seafarers to serve on their ships while the application for recognition is being processed.

**6.2.18**  If an endorsement to attest recognition or certificate of competency has expired or has not been issued or documentary proof of application for endorsement is not readily available, the PSCO should consider whether or not the ship can comply with STCW regulation I/4.1.2 regarding the numbers and certificates on board being in compliance with the applicable safe manning requirements of the flag State. This may be considered a deficiency in accordance with regulation I/4.2.4 and rectified before departure or detention may be applied. The officer carrying out the control should forthwith inform, in writing, the master of the ship and the Consul or, in his absence, the nearest diplomatic representative or the maritime authority of the State whose flag the ship is entitled to fly, so that appropriate action may be taken.

**6.2.19** In cases of suspected intoxication of masters, officers and/or other seafarers while performing designated safety, security and marine environmental protection duties, the appropriate Authorities of the port and flag State should be notified in accordance with chapters 3 and 4 of the Procedures for port State control.

**6.2.20** Seafarers should have a valid medical certificate and have completed applicable familiarization on board the ship. If such crew members are assigned to any designated safety, security or environmental prevention duties, they must be trained and qualified for such duties in accordance with the applicable chapter of the STCW Code.

**6.2.21** In accordance with section A-VI/1, paragraph 5 of the STCW Code, the flag State Administration may exempt the seafarers engaged on ships, other than passenger ships of more than 500 gross tonnage on international voyages and tankers from some of the requirements of that section.

### Manning

**6.2.22** The PSCO should examine the applicable documents, found in section 4.

**6.2.23** The guiding principles for port State control of the manning of a foreign ship should be:

    **.1** verification that the numbers and certificates of the seafarers serving on board are in conformity with the applicable safe manning requirements of the flag State; and

    **.2** verification that the vessel and its personnel conform to the international provisions as laid down in SOLAS and STCW.

**6.2.24** If a ship is manned in accordance with an MSMD or equivalent document issued by the flag State, the PSCO should accept that the ship is safely manned unless the document has clearly been issued without regard to the principles contained in the relevant instruments, in which case the PSCO should consult the flag State Administration.

**6.2.25** If the flag State Administration has not issued a safe manning document or equivalent due to the ship's size the PSCO should examine the CoC, CoP and their relevant flag State endorsement for the crew and compare with the requirements of the STCW Convention. Regarding the number of seafarers, the PSCO should then use his or her professional judgement, taking into account chapter VIII of the STCW Convention and Code and the duration and area of the next voyage, to determine if it can be undertaken safely. The PSCO should note the number of seafarers on board during the previous voyage as another indicator of standard manning levels for the ship. The PSCO should consult the flag State Administration, if additional information is necessary.

**6.2.26** If an endorsement to attest recognition has expired or has not been issued or documentary proof of application for endorsement (CRA) is not readily available, the PSCO should consider whether the ship can comply with the applicable safe manning requirements of the flag State Administration. In cases where the PSCO finds that additional information is necessary, the flag State Administration should be consulted.

**6.2.27** If the flag State does not respond to the request this should be considered as clear grounds for a more detailed inspection to ensure that the number and composition of the crew is in accordance with the principles laid down in paragraph 6.2.23 above. The ship should only be allowed to proceed to sea if it is safe to do so, taking into account the criteria for detention indicated in section 7.3. In any such case, the minimum standards to be applied should be no more stringent than those applied to ships flying the flag of the port State.

### Hours of rest

**6.2.28** All persons who are assigned duty as officer in charge of a watch or as a rating forming part of a watch and those whose duties involve designated safety, security and environmental protection duties shall be provided with a rest period of not less than:

    **.1**    minimum of 10 h of rest in any 24-hour period; and

    **.2**    77 h in any 7-day period.

**6.2.29** The hours of rest may be divided into no more than two periods, one of which shall be at least 6 h in length, and the intervals between consecutive periods of rest shall not exceed 14 h.

**6.2.30** The PSCO should examine the applicable documents, found in section 4, specifically the watch schedule and the records of daily hours of rest. The PSCO may inspect the seafarer's personal copy of his or her records pertaining to the hours of rest being held by the seafarer on board in order to verify that the records are accurate.

**6.2.31** The watch schedule shall be in a standardized format,* easily accessible to the crew and posted in the working language or languages of the ship and in English.

**6.2.32** Daily hours of rest shall be maintained in a standardized format,* in the working language or languages of the ship and in English.

**6.2.33** The PSCO should consider that seafarers who are on call, such as when a machinery space is unattended, are to be provided with an adequate compensatory rest period if the normal period is disturbed by call-outs to work.

**6.2.34** While assessing hours of rest, the PSCO should take into account any emergency conditions encountered which required a seafarer to perform additional hours of work for the immediate safety of the ship. In such cases, the master should be consulted for an explanation of the events and how impacted seafarers were provided with an adequate period of rest.

**6.2.35** Flag State Administrations may provide exceptions to the requirements of paragraphs 6.2.28.2 and 6.2.29 above for no more than two consecutive weeks provided that the rest period for the seafarer is not less than 70 h in any 7-day period.

## 6.3    Clear grounds

**6.3.1**    Clear grounds are defined in section 1.7.2 of the Procedures for port State control.

**6.3.2**    In addition to the general examples of clear grounds in section 2.4 of the Procedures, the specific occurrences below are considered as factors leading to a more detailed inspection:

    **.1**    the ship has been involved in a collision, grounding or stranding; or

    **.2**    there has been a discharge of substances from the ship when under way, at anchor or at berth which is illegal under any international convention; or

    **.3**    the ship has been manoeuvred in an erratic or unsafe manner whereby routeing measures adopted by IMO or safe navigation practices and procedures have not been followed; or

    **.4**    the ship is otherwise being operated in such a manner as to pose a danger to persons, property, or the environment, or a compromise to security; or

    **.5**    missing illegible or fraudulent certificates and records; or

    **.6**    flag State does not respond to requests for clarification of manning scales; or

---

\* The IMO/ILO Guidelines for the development of tables of seafarers' shipboard working arrangements and formats of records of seafarers' hours of work or hours of rest may be used.

.7  failure to conform to flag State requirements regarding watch arrangements (e.g. flag State requirements regarding certain ratings required to be on the bridge/in the engine-room during specific evolutions); or

.8  inability of crew member(s) to perform their assigned duties during abandon ship or fire-fighting drills; or

.9  inability of watchkeeping officer(s) to communicate with the PSCO in English; or

.10  inability of crew member(s) to operate shipboard equipment necessary to complete operational tests as required during the general examination; or

.11  clear indication, based on personal observations of performance during the inspection, that the master and/or crew are not familiar with their specific duties and with ship arrangements, installations, equipment, procedures and ship characteristics that are relevant to their routine or emergency duties; or

.12  indication that key crew members are not able to communicate or coordinate with each other, or with other persons on board; or

.13  failure to comply with the rest hour/fitness for duty provisions; or

.14  complaints received from a seafarer or knowledgeable party; or

.15  the ship has a master, officer or rating holding a certificate issued by a country which has not ratified the STCW Convention.

## 6.4  More detailed inspection

**6.4.1**  The PSCO should verify:

.1  that seafarers are sufficiently rested and otherwise fit for duty for the first watch at the commencement of the intended voyage and for subsequent relieving watches; this may be done by comparing records of daily hours of rest with the requirements in the STCW Convention for an appropriate period, which should at least include, whenever possible, the 7-day period immediately prior to departure; the rest period must reflect actual hours worked;

.2  a sufficient number of certificates from all departments to demonstrate that the vessel and the composition of the crew complies with the requirements of the STCW Convention; and

.3  that navigational or engineering watch arrangements conform to the requirements specified for the ship in the MSMD by the flag State and the requirements of STCW Convention regulation VIII/2, and STCW Code section A-VIII/2.

**6.4.2**  An assessment of seafarers can only be conducted by the port State if there are clear grounds for believing that the ability of the seafarers of the ship to maintain watchkeeping and security standards, as appropriate, as required by the STCW Convention is not being maintained because any of the situations mentioned in paragraphs 6.3.2.1 to 6.3.2.4 have occurred:

.1  The assessment procedure provided in the STCW Convention regulation I/4, paragraph 1.3, should take the form of a verification that members of the crew who are required to be competent do in fact possess the necessary skills related to the occurrence.

.2  It should be borne in mind when making this assessment that on-board procedures are relevant to the International Safety Management (ISM) Code and that the provisions of the STCW Convention are confined to the competence to safely execute those procedures.

.3  Control procedures under the STCW Convention should be confined to the standards of competence of the individual seafarers on board and their skills related to watchkeeping as defined in part A of the STCW Code. On-board assessment of competency should commence with verification of the certificates of the seafarers.

.4 Notwithstanding verification of the certificate, the assessment under the STCW Convention regulation I/4, paragraph 1.3 can require the seafarer to demonstrate the related competency at the place of duty. Such demonstration may include verification that operational requirements in respect of watchkeeping standards have been met and that there is a proper response to emergency situations within the seafarer's level of competence.

.5 In the assessment, only the methods for demonstrating competence together with the criteria for its evaluation and the scope of the standards given in part A of the STCW Code should be used. In cases where doubt of knowledge on operational use of equipment exists, the relevant officer or crew member should be asked to perform a functional test. Failure to perform a functional test could indicate the lack of familiarization or competency.

.6 Assessment of competency related to security should be conducted for those seafarers with specific security duties only in case of clear grounds, as provided for in chapter XI-2 of SOLAS, by the competent security Authority. In all other cases, it should be confined to the verification of the certificates and/or endorsements of the seafarers.

# 7 Follow-up action

## 7.1 Possible action

Possible action to be considered by the PSCO for the control in compliance with STCW or SOLAS Conventions may be dealt with in the following ways:

.1 exercise of control with regard to the documentation concerning the ship; and

.2 exercise of control with regard to the documentation for individual seafarers on board.

## 7.2 Possible deficiencies

The following is a non-exhaustive list of possible deficiencies:

*Seafarers' documentation:*

.1 no CoC, CoP, flag State endorsements or proof that an application for an endorsement has been submitted (STCW regulations I/4.2.1 and I/10);

.2 special training requirements: mandatory basic or advanced training or endorsement not presented;

.3 no evidence of basic training, or other certificate of proficiency, if not included in a qualification certificate held (STCW regulations VI/1, VI/1.2, VI/3, VI/4 and VI/6); and

.4 information or evidence that the master or crew is not familiar with essential shipboard operations relating to the safety of ships or the prevention of pollution, or that such operations have not been carried out.

*Manning:*

.5 no minimum safe manning document (MSMD) or the manning (number or qualification) not in accordance with the MSMD (STCW regulation I/4.2.2 and SOLAS regulation V/14); and

.6 unqualified person on duty (STCW regulation I/4.2.4).

*Hours of rest:*

.7 watch schedule not posted or not being followed (STCW regulations I/4.2.3 and I/4.2.5 and STCW Code A-VIII/1.5);

.8 the absence of a table of shipboard working arrangement or of records of rest of seafarers (STCW Code A-VIII/1.7);

.9 the records of hours of rest are inaccurate or incomplete (STCW Code A-VIII/1.7); and

.10 the watchkeeper is receiving less than 10 h rest in any 24-hour period (i.e. working in excess of 14 h) or 77 h rest in any 7-day period (STCW Code A-VIII/1).

## 7.3    Deficiencies warranting detention

**7.3.1**    A non-exhaustive list of grounds for detention is contained in regulation I/4 of the STCW Convention, as amended:

.1    failure of seafarers to hold a certificate, to have an appropriate certificate, to have a valid dispensation or to provide documentary proof that an application for an endorsement has been submitted to the Administration in accordance with regulation I/10, paragraph 5;

.2    failure to comply with the applicable safe manning requirement of the Administration;

.3    failure of navigational or engineering watch arrangements to conform to the requirements specified for the ship by the Administration;

.4    absence in a watch of a person qualified to operate equipment essential to safe navigation, safety radiocommunications or the prevention of marine pollution; and

.5    inability to provide, for the first watch at the commencement of a voyage and for subsequent relieving watches, persons who are sufficiently rested and otherwise fit for duty.

**7.3.2**    Other grounds for detention are listed below:

*Ship-related:*

.1    MSMD or equivalent not presented (SOLAS regulation V/14.2); and

.2    records of daily hours of rest are not on board (STCW Code A-VIII/1.7); and

*Seafarers' documentation:*

.3    not available or serious discrepancy in the CoC (STCW regulation I/4.2.1);

.4    absence in watch of a radio operator (general/restricted GMDSS); certificates and endorsement not available (STCW regulations I/4.2.1, I/4.2.2, I/4.2.3, I/4.2.4 and II/1.2.1);

.5    documentation for personnel with designated safety, security and marine environmental duties not available (STCW regulation I/4.2.1, I/4.2.2, I/4.2.3 and I/4.2.4);

.6    expired certificates (STCW regulation I/4.2.1), and for medical certificates also refer to STCW regulation I/9.6 and I/9.7; and

.7    evidence that a certificate has been fraudulently obtained or the holder of a certificate is not the person to whom that certificate was originally issued.

## 7.4    Actions to be considered

### Ship-related

**7.4.1**    If the actual number of crew or composition does not conform to the manning document, the port State should request the flag State for advice as to whether or not the ship should be allowed to sail with the actual number of crew and composition of crew. Such a request and response should be by the most expedient means and either party may request the communication in writing. If the actual crew number or composition is not brought into compliance with the MSMD or the flag State does not advise that the ship may sail, the ship may be considered for detention after the criteria set out in section 7.3 have been taken into account.

**7.4.2**    Before detaining the ship the PSCO should consider the following:

.1    length and nature of the intended voyage or service;

.2    whether or not the deficiency poses a danger to ships, persons on board or the environment;

.3    whether or not appropriate rest periods of the crew can be observed;

.4    size and type of ship and equipment provided; and

.5    nature of cargo.

### Deficiency-related

**7.4.3**   When the manning is not in accordance with the MSMD and no flag State endorsements or no "documentary proof of application" can be presented, the port State should consult the flag State whenever possible taking into account time differences or other conditions. However, if it is not possible to establish contact with the flag State, the port State should forthwith inform, in writing, the master of the ship and the Consul or, in their absence, the nearest diplomatic representative or the maritime authority of the State whose flag the ship is entitled to fly, so that appropriate action may be taken.

**7.4.4**   In cases where an unqualified seafarer has been on duty and/or the watch schedule has not been followed, the flag State should be informed and this could be considered as an ISM deficiency.

**7.4.5**   In cases where there is a seafarer on duty who is not qualified to carry out an operation, that particular operation should be stopped immediately.

## 8    Note on reporting deficiencies

The PSCO should be aware that, in addition to SOLAS and STCW, there may be other applicable international instruments. The PSCO should decide which one is the most appropriate.

## Annex
*Table B-I/2*
## *List of certificates or documentary evidence required under the STCW Convention*

The list below identifies all certificates or documentary evidence described in the STCW Convention which authorize the holder to serve in certain functions on board ships. The certificates are subject to the requirements of regulation I/2 regarding language and their availability in original form.

The list also references the relevant regulations and the requirements for endorsement, registration and revalidation.

| STCW Regulation | Type of certificate | Endorsement attesting recognition[1] | Registration[2] | Revalidation[3] |
|---|---|---|---|---|
| II/1, II/2, II/3, III/1, III/2, III/3, III/6, IV/2, VII/2 | Certificate of Competency – for masters, officers and GMDSS radio operators | Yes | Yes | Yes |
| II/4, III/4, VII/2 | Certificate of Proficiency – for ratings duly certified to be part of a navigational or engine-room watch | No | Yes | No |
| II/5, III/5, III/7, VII/2 | Certificate of Proficiency – for ratings duly certified as able seafarer deck, able seafarer engine or electro-technical rating | No | Yes | No |
| V/1-1, V/1-2 | Certificate of Proficiency or endorsement to a Certificate of Competency – for masters and officers on oil, chemical or liquefied gas tankers | Yes | Yes | Yes |
| V/1-1, V/1-2 | Certificate of Proficiency – for ratings on oil, chemical or liquefied gas tankers | No | Yes | No |
| V/2 | Documentary evidence – Training for masters, officers, ratings and other personnel serving on passenger ships | No | No | No[4] |

| STCW Regulation | Type of certificate | Endorsement attesting recognition[1] | Registration[2] | Revalidation[3] |
|---|---|---|---|---|
| V/3 | Certificate of Proficiency – training for masters, officers, ratings and other personnel on ships subject to the IGF Code | No | Yes | Yes[8] |
| V/4 | Certificate of Proficiency – for masters and officers on ships operating in polar waters | No | Yes | Yes |
| VI/1 | Certificate of Proficiency[5] – Basic training | No | Yes | Yes[6] |
| VI/2 | Certificate of Proficiency[5] – Survival craft, rescue boats and fast rescue boats | No | Yes | Yes[6] |
| VI/3 | Certificate of Proficiency[5] – Advanced fire fighting | No | Yes | Yes[6] |
| VI/4 | Certificate of Proficiency[5] – Medical first aid and medical care | No | Yes | No |
| VI/5 | Certificate of Proficiency – Ship security officer | No | Yes | No |
| VI/6 | Certificate of Proficiency[7] – security awareness training or security training for seafarers with designated security duties | No | Yes | No |

## Notes

**1** *Endorsement attesting recognition of a certificate* means endorsement in accordance with regulation I/2, paragraph 7.

**2** *Registration required* means as part of register or registers in accordance with regulation I/2, paragraph 14.

**3** *Revalidation of a certificate* means establishing continued professional competence in accordance with regulation I/11 or maintaining the required standards of competence in accordance with sections A-V/3 and A-VI/1 to A-VI/3, as applicable.

**4** As required by regulation V/2, paragraph 4 seafarers who have completed training in "crowd management", "crisis management and human behaviour" or "passenger safety, cargo safety and hull integrity" shall at intervals not exceeding five years, undertake appropriate refresher training or to provide evidence of having achieved the required standards of competence within the previous five years.

**5** The certificates of competency issued in accordance with regulations II/1, II/2, II/3, III/1, III/2, III/3, III/6 and VII/2 include the proficiency requirements in "basic training", "survival craft and rescue boats other than fast rescue boats", "advanced fire fighting" and "medical first aid" therefore, holders of mentioned certificates of competency are not required to carry Certificates of Proficiency in respect of those competences of chapter VI.

**6** In accordance with sections A-VI/1, A-VI/2 and A-VI/3, seafarers shall provide evidence of having maintained the required standards of competence every five years.

**7** Where security awareness training or training in designated security duties is not included in the qualification for the certificate to be issued.

**8** In accordance with regulation V/3, seafarers shall, at intervals not exceeding five years, undertake appropriate refresher training or be required to provide evidence of having achieved the required standard of competence within the previous five years.

# Appendix 12
## *List of certificates and documents*

List of certificates and documents which to the extent applicable should be checked during the inspection referred to in paragraph 2.2.3 (as appropriate):

1      International Tonnage Certificate (1969);

2      Reports of previous port State control inspections;

3      Passenger Ship Safety Certificate (SOLAS reg.I/12);

4      Cargo Ship Safety Construction Certificate (SOLAS reg.I/12);

5      Cargo Ship Safety Equipment Certificate (SOLAS reg.I/12);

6      Cargo Ship Safety Radio Certificate (SOLAS reg.I/12);

7      Cargo Ship Safety Certificate (SOLAS reg.I/12);

8      For ro-ro passenger ships, information on the A/A-max ratio (SOLAS reg.II-1/8-1[*]);

9      Damage control plans and booklets (SOLAS reg.II-1/19);

10     Stability information (SOLAS regs.II-1/5 and II-1/5-1 and LLC 66/88 reg.10);

11     Manoeuvring Booklet and information (SOLAS reg.II-1/28);

12     Unattended machinery spaces (UMS) evidence (SOLAS reg.II-I/46.3);

13     Fixed gas fire-extinguishing systems – cargo spaces Exemption Certificate and any list of cargoes (SOLAS reg.II-2/10.7.1.4);

14     Fire control plan/booklet (SOLAS reg.II-2/15.2.4 and II-2/15.3.2);

15     Fire safety operational booklet (SOLAS reg.II-2/16.2);

16     Dangerous goods manifest or stowage plan (SOLAS reg.VII/4 and VII/7-2; MARPOL Annex III reg.5);

17     Document of compliance with the special requirements for ships carrying dangerous goods (SOLAS reg.II-2/19.4);

18     Onboard training, drills and maintenance records (SOLAS reg.II-2/15.2.2.5 and reg.III/19.3 and III/19.5 and III/20.6 and III/20.7);

19     Minimum safe manning document (SOLAS reg.V/14.2);

20     Search and Rescue cooperation plan for passenger ships trading on fixed routes (SOLAS reg.V/7.3);

21     LRIT Conformance Test Report (SOLAS reg.V/19-1.6);

22     Copy of the Certificate of compliance issued by the testing facility, stating the date of compliance and the applicable performance standards of VDR (voyage data recorder) (SOLAS reg.V/18.8);

23     For passenger ships, List of operational limitations (SOLAS reg.V/30.2);

24     Cargo Securing Manual (SOLAS reg.VI/5.6 and VII/5; MSC.1/Circ.1353);

25     Bulk Carrier Booklet (SOLAS reg.VI/7.2 and XII/8);

26     Loading/Unloading Plan for bulk carriers (SOLAS reg.VI/7.3);

---

[*] Refer to Resolution 1 of the 1995 SOLAS Conference.

27    Document of authorization for the carriage of grain and grain loading manual (SOLAS reg.VI/9; International Code for the Safe Carriage of Grain in Bulk, section 3);

28    INF (International Code for the Safe Carriage of Packaged Irradiated Nuclear Fuel, Plutonium and High-Level Radioactive Wastes on Board Ships) Certificate of Fitness (SOLAS reg.VII/16 and INF Code reg.1.3);

29    Copy of Document of Compliance issued in accordance with the International Management Code for the Safe Operation of Ships and for Pollution Prevention (ISM Code) (DoC) (SOLAS reg.IX/4.2, ISM Code, paragraph 13);

30    Safety Management Certificate issued in accordance with the International Management Code for the Safe Operation of Ships and for Pollution Prevention (SMC) (SOLAS reg.IX/4.3, ISM Code, paragraph 13);

31    High-Speed Craft Safety Certificate and Permit to Operate High-Speed Craft (SOLAS reg.X/3.2 and HSC Code 94/00 reg.1.8.1 and 1.9);

32    Continuous Synopsis Record (SOLAS reg.XI-1/5);

33    International Certificate of Fitness for the Carriage of Liquefied Gases in Bulk, or the Certificate of Fitness for the Carriage of Liquefied Gases in Bulk, whichever is appropriate (IGC Code reg.1.5.4 or GC Code reg.1.6);

34    International Certificate of Fitness for the Carriage of Dangerous Chemicals in Bulk, or the Certificate of Fitness for the Carriage of Dangerous Chemicals in Bulk, whichever is appropriate (IBC Code reg.1.5.4 and BCH Code reg.1.6.3);

35    International Oil Pollution Prevention Certificate (MARPOL Annex I reg.7.1);

36    Enhanced Survey Report Files (in case of bulk carriers or oil tankers) (SOLAS reg.XI-1/2 and 2011 ESP Code paragraphs 6.2 and 6.3 of annex A, part A and part B, and annex B, part A and part B);

37    Oil Record Book, parts I and II (MARPOL Annex I regs.17 and 36);

38    Shipboard Marine Pollution Emergency Plan for Noxious Liquid Substances (MARPOL Annex II reg.17);

39    Statement of compliance Condition Assessment Scheme (CAS) (MARPOL Annex I regs.20.6 and 21.6.1);

40    For oil tankers, the record of oil discharge monitoring and control system for the last ballast voyage (MARPOL Annex I reg.31.2);

41    Shipboard Oil Pollution Emergency Plan (MARPOL Annex I reg.37.1);

42    International Pollution Prevention Certificate for the Carriage of Noxious Liquid Substances in Bulk (NLS) (MARPOL Annex II reg.9.1);

43    Cargo Record Book (MARPOL Annex II reg.15);

44    Procedures and Arrangements Manual (chemical tankers) (MARPOL Annex II reg.14.1);

45    International Sewage Pollution Prevention Certificate (ISPPC) (MARPOL Annex IV reg.5.1);

46    Garbage Management Plan (MARPOL Annex V reg.10);

47    Garbage Record Book (MARPOL Annex V reg.10);

48    International Air Pollution Prevention Certificate (IAPPC) (MARPOL Annex VI reg.6.1);

49    Fuel oil Changeover Procedure and Logbook for fuel oil changeover (MARPOL Annex VI reg.14.6);

50    Type approval certificate of incinerator (MARPOL Annex VI reg.16.6);

51    Bunker delivery notes and Representative Sample (MARPOL Annex VI reg.18.6 and 18.8.1);

52    Engine International Air Pollution Prevention Certificate (EIAPPC) (NOX Technical Code 2008 reg.2.1.1.1);

53    Technical files (NOX Technical Code 2008 reg.2.3.4);

54    Record book of engine parameters (NOX Technical Code reg.2.3.7);

55    International Load Line Certificate (1966) (LLC 66/88 art.16.1);

56    International Load Line Exemption Certificate (LLC 66/88 art.16.2);

57    Certificates for masters, officers or ratings issued in accordance with STCW Convention (STCW art.VI, reg.I/2 and STCW Code section A-I/2);

58    Records of hours of rest and table of shipboard working arrangements (STCW Code section A-VIII/1.5 and 1.7, ILO Convention No.180 art.5.7, art.8.1 and MLC, 2006 Standard A.2.3.10 and A.2.3.12);

59    Certificate of insurance or any other financial security in respect of civil liability for oil pollution damage (CLC 69/92 art.VII.2);

60    Certificate of insurance or any other financial security in respect of civil liability for bunker oil pollution damage (BUNKERS 2001 art.7.2);

61    International Ship Security Certificate (ISSC) or Interim International Ship Security Certificate (ISPS Code part A/19 and appendices);

62    Record of AFS (AFS 2001 Annex 4 reg.2);

63    International Anti-fouling System Certificate (IAFS Certificate) (AFS 2001 Annex 4 reg.2); and

64    Declaration on AFS (AFS 2001 Annex 4 reg.5).

65    Coating Technical File (SOLAS reg.II-1/3-2);

66    Construction drawings (SOLAS reg.II-1/3-7);

67    Ship Construction File (SOLAS reg.II-1/3-10);

68    Fire safety training manual (SOLAS reg.II-2/1 5.2.3);

69    Maintenance plans (SOLAS reg.II-2/14.2.2 and II-2/14.4);

70    Training manual (SOLAS reg.III/35);

71    Nautical charts and nautical publications (SOLAS reg.V/19.2.1.4 and V/27);

72    International Code of Signals and a copy of Volume III of IAMSAR Manual (SOLAS reg.V/21);

73    Records of navigational activities (SOLAS reg.V/26 and V/28.1);

74    Material Safety Data Sheets (MSDS) (SOLAS reg.VI/5-1);

75    AIS test report (SOLAS reg.V/18.9);

76    Ship Security Plan and associated records (SOLAS reg.XI-2/9 and ISPS Code part A/9 and 10);

77    International Energy Efficiency Certificate (MARPOL Annex VI reg.6);

78    Ozone-depleting Substances Record Book (MARPOL Annex VI reg.12.6);

79    Manufacturer's Operating Manual for Incinerators (MARPOL Annex VI reg.16.7);

80    Ship Energy Efficiency Management Plan (MARPOL Annex VI reg.22);

81    EEDI Technical File (MARPOL Annex VI, reg.20);

82    Noise Survey Report (SOLAS reg.II-1/3-12);

83    Ship-specific Plans and Procedures for Recovery of Persons from the Water (SOLAS reg.III/17-1);

84    Decision support system for masters (Passenger ships) (SOLAS reg.III/29);

85    Oil Discharge Monitoring and Control (ODMC) Operational Manual (MARPOL Annex I reg.31);

86    Cargo Information (SOLAS reg.VI/2 and XII/10);

87    Ship Structure Access Manual (SOLAS reg.II-1/3-6);

88    Crude Oil Washing Operation and Equipment Manual (MARPOL Annex I reg.35);

89    Subdivision and stability information (MARPOL Annex I reg.28);

90    STS Operation Plan and Records of STS Operations (MARPOL Annex I reg.41);

91    VOC Management Plan (MARPOL Annex VI reg.15.6);

92    Exemption Certificate (SOLAS reg.I/12);

**93** Certificate of Insurance or other Financial Security in respect of Liability for the Removal of Wrecks (Removal of Wreck Article 12);

**94** International Ballast Water Management Certificate (IBWMC) (BWMC Art 9.1(a) and reg.E-2);

**95** Ballast Water Record Book (BWRB) (BWMC Art 9.1 (b) and reg.B-2).

**96** Ballast Water Management Plan (BWMP) (BWMC reg.B-1)

**For reference**

**1** Certificate of Registry or other document of nationality (UNCLOS art.9.1.2);

**2** Certificates as to the ship's hull strength and machinery installations issued by the classification society in question (only to be required if the ship maintains its class with a classification society);

**3** Cargo Gear Record Book (ILO Convention No.32 art.9.2(4) and ILO Convention No.152 art.25);

**4** Certificates loading and unloading equipment (ILO Convention No.134 art.4.3(e) and ILO Convention No.32 art.9(4));

**5** Medical certificates (ILO Convention No.73 or MLC, 2006 Standard A1.2);

**6** Records of hours of work or rest of seafarers (ILO Convention No.180 part II art.8.1 or MLC, 2006, Standard A.2.3.12);

**7** Maritime Labour Certificate (MLC, 2006, Regulation 5.1.3);

**8** Declaration of Maritime Labour compliance (DMLC) on board (parts I and II) (MLC, 2006, Regulation 5.1.3);

**9** Seafarer's employment agreements (MLC, 2006, Standard A 2.1);

**10** Certificate of Insurance or Financial Security for Repatriation of Seafarers (MLC, 2006, Regulation 2.5); and

**11** Certificate of Insurance or Financial Security for Shipowners liability (MLC, 2006, Regulation 4.2).

# Appendix 13

## REPORT OF INSPECTION IN ACCORDANCE WITH IMO PORT STATE CONTROL PROCEDURES[*]

### FORM A

(Reporting authority)                                Copy to:    Master

(Address)                                                        Head office

(Telephone)                                                      PSCO

(Telefax)

(Email)

If ship is detained, copy to:
Flag State
IMO
Recognized organization, if applicable

1    Name of reporting authority . . . . . . . . . . . . . . . . . . . . . . . . . . . . . . . . . . . . . . . . . . . . . . . . . . . . . . . .

2    Name of ship . . . . . . . . . . . . . . . . . . . . . . . . . . . . . . . . . . . . . . . . . . . . . . . . . . . . . . . . . . . . . . . . . . . .

3    Flag of ship . . . . . . . . . . . . . . . . . . . . . . . . . . . . . . . . . . . . . . . . . . . . . . . . . . . . . . . . . . . . . . . . . . . . .

4    Type of ship . . . . . . . . . . . . . . . . . . . . . . . . . . . . . . . . . . . . . . . . . . . . . . . . . . . . . . . . . . . . . . . . . . . .

5    Call sign . . . . . . . . . . . . . . . . . . . . . . . . . . . . . . . . . . . . . . . . . . . . . . . . . . . . . . . . . . . . . . . . . . . . . .

6    IMO number . . . . . . . . . . . . . . . . . . . . . . . . . . . . . . . . . . . . . . . . . . . . . . . . . . . . . . . . . . . . . . . . . . .

7    Gross tonnage . . . . . . . . . . . . . . . . . . . . . . . . . . . . . . . . . . . . . . . . . . . . . . . . . . . . . . . . . . . . . . . . .

8    Deadweight (where applicable) . . . . . . . . . . . . . . . . . . . . . . . . . . . . . . . . . . . . . . . . . . . . . . . . . . . . .

9    Year of build . . . . . . . . . . . . . . . . . . . . . . . . . . . . . . . . . . . . . . . . . . . . . . . . . . . . . . . . . . . . . . . . . .

10  Date of inspection . . . . . . . . . . . . . . . . . . . . . . . . . . . . . . . . . . . . . . . . . . . . . . . . . . . . . . . . . . . . . .

11  Place of inspection . . . . . . . . . . . . . . . . . . . . . . . . . . . . . . . . . . . . . . . . . . . . . . . . . . . . . . . . . . . . .

12  Classification society . . . . . . . . . . . . . . . . . . . . . . . . . . . . . . . . . . . . . . . . . . . . . . . . . . . . . . . . . . .

13  Date of release from detention[†] . . . . . . . . . . . . . . . . . . . . . . . . . . . . . . . . . . . . . . . . . . . . . . . . . .

14  Particulars of ISM company (details or IMO Company Number)[†] . . . . . . . . . . . . . . . . . . . . . . . . . . .

15  Relevant certificate(s)[†] . . . . . . . . . . . . . . . . . . . . . . . . . . . . . . . . . . . . . . . . . . . . . . . . . . . . . . . . . .

| | a) Title | b) Issuing authority | c) Dates of issue and expiry |
|---|---|---|---|
| 1 | . . . . . . . . . . . . . . . . . . . . . | . . . . . . . . . . . . . . . . . . . . . . . . | . . . . . . . . . . . . . . . . . . . . . . . . |
| 2 | . . . . . . . . . . . . . . . . . . . . . | . . . . . . . . . . . . . . . . . . . . . . . . | . . . . . . . . . . . . . . . . . . . . . . . . |
| 3 | . . . . . . . . . . . . . . . . . . . . . | . . . . . . . . . . . . . . . . . . . . . . . . | . . . . . . . . . . . . . . . . . . . . . . . . |
| 4 | . . . . . . . . . . . . . . . . . . . . . | . . . . . . . . . . . . . . . . . . . . . . . . | . . . . . . . . . . . . . . . . . . . . . . . . |
| 5 | . . . . . . . . . . . . . . . . . . . . . | . . . . . . . . . . . . . . . . . . . . . . . . | . . . . . . . . . . . . . . . . . . . . . . . . |
| 6 | . . . . . . . . . . . . . . . . . . . . . | . . . . . . . . . . . . . . . . . . . . . . . . | . . . . . . . . . . . . . . . . . . . . . . . . |

---

[*] This inspection report has been issued solely for the purposes of informing the master and other port States that an inspection by the port State, mentioned in the heading, has taken place. This inspection report cannot be construed as a seaworthiness certificate **in excess of** the certificate the ship is required to carry.

[†] To be completed in the event of a detention.

| 7 | . . . . . . . . . . . . . . . . . . . . . . . . . . | . . . . . . . . . . . . . . . . . . . . . . . . . . | . . . . . . . . . . . . . . . . . . . . . . . . . . |
| 8 | . . . . . . . . . . . . . . . . . . . . . . . . . . | . . . . . . . . . . . . . . . . . . . . . . . . . . | . . . . . . . . . . . . . . . . . . . . . . . . . . |
| 9 | . . . . . . . . . . . . . . . . . . . . . . . . . . | . . . . . . . . . . . . . . . . . . . . . . . . . . | . . . . . . . . . . . . . . . . . . . . . . . . . . |
| 10 | . . . . . . . . . . . . . . . . . . . . . . . . . . | . . . . . . . . . . . . . . . . . . . . . . . . . . | . . . . . . . . . . . . . . . . . . . . . . . . . . |
| 11 | . . . . . . . . . . . . . . . . . . . . . . . . . . | . . . . . . . . . . . . . . . . . . . . . . . . . . | . . . . . . . . . . . . . . . . . . . . . . . . . . |
| 12 | . . . . . . . . . . . . . . . . . . . . . . . . . . | . . . . . . . . . . . . . . . . . . . . . . . . . . | . . . . . . . . . . . . . . . . . . . . . . . . . . |

d). . . . . . . . . . . . . . . . . . . . . . . . . . . . . . . . . . . . . . . . Information on last intermediate or annual survey[*]

| | Date | Surveying authority | Place |
|---|---|---|---|
| 1 | . . . . . . . . . . . . . . . . . . . . . . . . . . | . . . . . . . . . . . . . . . . . . . . . . . . . . | . . . . . . . . . . . . . . . . . . . . . . . . . . |
| 2 | . . . . . . . . . . . . . . . . . . . . . . . . . . | . . . . . . . . . . . . . . . . . . . . . . . . . . | . . . . . . . . . . . . . . . . . . . . . . . . . . |
| 3 | . . . . . . . . . . . . . . . . . . . . . . . . . . | . . . . . . . . . . . . . . . . . . . . . . . . . . | . . . . . . . . . . . . . . . . . . . . . . . . . . |
| 4 | . . . . . . . . . . . . . . . . . . . . . . . . . . | . . . . . . . . . . . . . . . . . . . . . . . . . . | . . . . . . . . . . . . . . . . . . . . . . . . . . |
| 5 | . . . . . . . . . . . . . . . . . . . . . . . . . . | . . . . . . . . . . . . . . . . . . . . . . . . . . | . . . . . . . . . . . . . . . . . . . . . . . . . . |
| 6 | . . . . . . . . . . . . . . . . . . . . . . . . . . | . . . . . . . . . . . . . . . . . . . . . . . . . . | . . . . . . . . . . . . . . . . . . . . . . . . . . |
| 7 | . . . . . . . . . . . . . . . . . . . . . . . . . . | . . . . . . . . . . . . . . . . . . . . . . . . . . | . . . . . . . . . . . . . . . . . . . . . . . . . . |
| 8 | . . . . . . . . . . . . . . . . . . . . . . . . . . | . . . . . . . . . . . . . . . . . . . . . . . . . . | . . . . . . . . . . . . . . . . . . . . . . . . . . |
| 9 | . . . . . . . . . . . . . . . . . . . . . . . . . . | . . . . . . . . . . . . . . . . . . . . . . . . . . | . . . . . . . . . . . . . . . . . . . . . . . . . . |
| 10 | . . . . . . . . . . . . . . . . . . . . . . . . . . | . . . . . . . . . . . . . . . . . . . . . . . . . . | . . . . . . . . . . . . . . . . . . . . . . . . . . |
| 11 | . . . . . . . . . . . . . . . . . . . . . . . . . . | . . . . . . . . . . . . . . . . . . . . . . . . . . | . . . . . . . . . . . . . . . . . . . . . . . . . . |
| 12 | . . . . . . . . . . . . . . . . . . . . . . . . . . | . . . . . . . . . . . . . . . . . . . . . . . . . . | . . . . . . . . . . . . . . . . . . . . . . . . . . |

| 16 | Deficiencies | ☐ No | ☐ Yes (see attached FORM B) |
|---|---|---|---|
| 17 | Penalty imposed | ☐ No | ☐ Yes    Amount: |
| 18 | Ship detained | ☐ No | ☐ Yes[*] |
| 19 | Supporting documentation | ☐ No | ☐ Yes (see annex) |

Issuing office . . . . . . . . . . . . . . . . . . . . . . . . . . . . .     Name . . . . . . . . . . . . . . . . . . . . . . . . . . . . . . . . . . . .

*(duly authorized PSCO
of reporting authority)*

Telephone . . . . . . . . . . . . . . . . . . . . . . . . . . . . . .

Telefax . . . . . . . . . . . . . . . . . . . . . . . . . . . . . . .     Signature . . . . . . . . . . . . . . . . . . . . . . . . . . . . . . . .

**This report must be retained on board for a period of two years and must be available for consultation by port State control officers at all times.**

---

[*] Masters, shipowners and/or operators are advised that detailed information on a detention may be subject to future publication.

# REPORT OF INSPECTION IN ACCORDANCE WITH
## IMO PORT STATE CONTROL PROCEDURES

### FORM B

(Reporting authority)         Copy to:     Master

(Address)                                        Head office

(Telephone)                                 PSCO

(Telefax)

(Email)

If ship is detained, copy to:
Flag State
IMO
Recognized organization, if applicable

2      Name of ship . . . . . . . . . . . . . . . . . . . . . . . . . . . . . . . . . . . . . . . . . . . . . . . . . . . . . . . . . . . . . . .

6      IMO number . . . . . . . . . . . . . . . . . . . . . . . . . . . . . . . . . . . . . . . . . . . . . . . . . . . . . . . . . . . . . . .

10     Date of inspection . . . . . . . . . . . . . . . . . . . . . . . . . . . . . . . . . . . . . . . . . . . . . . . . . . . . . . . . .

11     Place of inspection . . . . . . . . . . . . . . . . . . . . . . . . . . . . . . . . . . . . . . . . . . . . . . . . . . . . . . . .

20. . . . . . . . . . . . . . . . . . . . . . . . . . .                              21. . . . . . . . . . . . . . . . . . . . . . . . . . .

Nature of deficiency[*]             Convention[†]                   Action taken[‡]

| Nature of deficiency | Convention | Action taken |
|---|---|---|
| . . . . . . . . . . . | . . . . . . . . . . . | . . . . . . . . . . . |
| . . . . . . . . . . . | . . . . . . . . . . . | . . . . . . . . . . . |
| . . . . . . . . . . . | . . . . . . . . . . . | . . . . . . . . . . . |
| . . . . . . . . . . . | . . . . . . . . . . . | . . . . . . . . . . . |
| . . . . . . . . . . . | . . . . . . . . . . . | . . . . . . . . . . . |
| . . . . . . . . . . . | . . . . . . . . . . . | . . . . . . . . . . . |
| . . . . . . . . . . . | . . . . . . . . . . . | . . . . . . . . . . . |

Name . . . . . . . . . . . . . . . . . . . . . . . . . . . . . . . . . . . . .

*(duly authorized PSCO
of reporting authority)*

Signature . . . . . . . . . . . . . . . . . . . . . . . . . . . . . . . . . . . .

---

[*] This inspection was not a full survey and deficiencies listed may not be exhaustive. In the event of a detention, it is recommended that full survey is carried out and all deficiencies are rectified before an application for re-inspection is made.

[†] To be completed in the event of a detention.

[‡] Actions taken include, e.g., ship detained/released, flag State informed, classification society informed, next port informed.

# Appendix 14

In accordance with the provision of paragraph 3.7.3 of
IMO Port State Control Procedures (resolution A.1119(30))

(Copy to maritime Authority of next port of call, flag Administration,
or other certifying authority as appropriate)

1    From (country/region) . . . . . . . . . . . . . . . . . . . . . . . . . . . . . . . . . . . . . . . . . . . . . . . . . . . . . . . . . . . .

2    Port . . . . . . . . . . . . . . . . . . . . . . . . . . . . . . . . . . . . . . . . . . . . . . . . . . . . . . . . . . . . . . . . . . . . . . . . . .

3    To (country/region). . . . . . . . . . . . . . . . . . . . . . . . . . . . . . . . . . . . . . . . . . . . . . . . . . . . . . . . . . . . . .

4    Port . . . . . . . . . . . . . . . . . . . . . . . . . . . . . . . . . . . . . . . . . . . . . . . . . . . . . . . . . . . . . . . . . . . . . . . . . .

5    Name of ship . . . . . . . . . . . . . . . . . . . . . . . . . . . . . . . . . . . . . . . . . . . . . . . . . . . . . . . . . . . . . . . . . . .

6    Date departed . . . . . . . . . . . . . . . . . . . . . . . . . . . . . . . . . . . . . . . . . . . . . . . . . . . . . . . . . . . . . . . . . .

7    Estimated place and time of arrival. . . . . . . . . . . . . . . . . . . . . . . . . . . . . . . . . . . . . . . . . . . . . . . . . . .

8    IMO number. . . . . . . . . . . . . . . . . . . . . . . . . . . . . . . . . . . . . . . . . . . . . . . . . . . . . . . . . . . . . . . . . . . .

9    Flag of ship and POR . . . . . . . . . . . . . . . . . . . . . . . . . . . . . . . . . . . . . . . . . . . . . . . . . . . . . . . . . . . . .

10   Type of ship . . . . . . . . . . . . . . . . . . . . . . . . . . . . . . . . . . . . . . . . . . . . . . . . . . . . . . . . . . . . . . . . . . . .

11   Call sign. . . . . . . . . . . . . . . . . . . . . . . . . . . . . . . . . . . . . . . . . . . . . . . . . . . . . . . . . . . . . . . . . . . . . . . .

12   Gross tonnage. . . . . . . . . . . . . . . . . . . . . . . . . . . . . . . . . . . . . . . . . . . . . . . . . . . . . . . . . . . . . . . . . . .

13   Year of build. . . . . . . . . . . . . . . . . . . . . . . . . . . . . . . . . . . . . . . . . . . . . . . . . . . . . . . . . . . . . . . . . . . .

14   Issuing authority of relevant certificate(s) . . . . . . . . . . . . . . . . . . . . . . . . . . . . . . . . . . . . . . . . . . . . . .

15   Nature of deficiencies to be rectified        16   Suggested action (including action at next port of call)

| | |
|---|---|
| . . . . . . . . . . . . . . . . . . . . . . . . . . . . . . . . . . . . . . | . . . . . . . . . . . . . . . . . . . . . . . . . . . . . . . . . . . . . . |
| . . . . . . . . . . . . . . . . . . . . . . . . . . . . . . . . . . . . . . | . . . . . . . . . . . . . . . . . . . . . . . . . . . . . . . . . . . . . . |
| . . . . . . . . . . . . . . . . . . . . . . . . . . . . . . . . . . . . . . | . . . . . . . . . . . . . . . . . . . . . . . . . . . . . . . . . . . . . . |
| . . . . . . . . . . . . . . . . . . . . . . . . . . . . . . . . . . . . . . | . . . . . . . . . . . . . . . . . . . . . . . . . . . . . . . . . . . . . . |
| . . . . . . . . . . . . . . . . . . . . . . . . . . . . . . . . . . . . . . | . . . . . . . . . . . . . . . . . . . . . . . . . . . . . . . . . . . . . . |

17   Action taken

. . . . . . . . . . . . . . . . . . . . . . . . . . . . . . . . . . . . . . . . . . . . . . . . . . . . . . . . . . . . . . . . . . . . . . . . . . . . . . . .

. . . . . . . . . . . . . . . . . . . . . . . . . . . . . . . . . . . . . . . . . . . . . . . . . . . . . . . . . . . . . . . . . . . . . . . . . . . . . . . .

Reporting Authority . . . . . . . . . . . . . . . . . . . . . . . . . . . . Name . . . . . . . . . . . . . . . . . . . . . . . . . . . . . . . . . . . . . . . . .
*(duly authorized PSCO
of reporting authority)*

Office. . . . . . . . . . . . . . . . . . . . . . . . . . . . . . . . . . . . . . . . Signature . . . . . . . . . . . . . . . . . . . . . . . . . . . . . . . . . . . .

Telefax/Email. . . . . . . . . . . . . . . . . . . . . . . . . . . . . . . . . Date . . . . . . . . . . . . . . . . . . . . . . . . . . . . . . . . . . . . . . . .

# Appendix 15

## REPORT OF ACTION TAKEN TO THE NOTIFYING AUTHORITY
### In accordance with the provision of paragraph 3.7.3 of
### IMO port State control procedures (resolution A.1119(30))
#### (by telefax/email and/or mail)

1  To: (Name). . . . . . . . . . . . . . . . . . . . . . . . . . . . . . . . . . . . . . . . . . . . . . . . . . . . . . . . . . . . . . . . . . . . . . . . . .

(Position) . . . . . . . . . . . . . . . . . . . . . . . . . . . . . . . . . . . . . . . . . . . . . . . . . . . . . . . . . . . . . . . . . . . . . . .

(Authority) . . . . . . . . . . . . . . . . . . . . . . . . . . . . . . . . . . . . . . . . . . . . . . . . . . . . . . . . . . . . . . . . . . . . . .

Telephone . . . . . . . . . . . . . . . . . . . . . . . . . . . . . . . . . . Telefax/Email . . . . . . . . . . . . . . . . . . . . . . . . . . .

Date. . . . . . . . . . . . . . . . . . . . . . . . . . . . . . . . . . . . . . . . . . . . . . . . . . . . . . . . . . . . . . . . . . . . . . . . . . . . . .

2  To: (Name) . . . . . . . . . . . . . . . . . . . . . . . . . . . . . . . . . . . . . . . . . . . . . . . . . . . . . . . . . . . . . . . . . . . . . . . .

(Position) . . . . . . . . . . . . . . . . . . . . . . . . . . . . . . . . . . . . . . . . . . . . . . . . . . . . . . . . . . . . . . . . . . . . . . .

(Authority) . . . . . . . . . . . . . . . . . . . . . . . . . . . . . . . . . . . . . . . . . . . . . . . . . . . . . . . . . . . . . . . . . . . . . .

Telephone . . . . . . . . . . . . . . . . . . . . . . . . . . . . . . . . . Telefax/Email . . . . . . . . . . . . . . . . . . . . . . . . . . .

3  Name of ship . . . . . . . . . . . . . . . . . . . . . . . . . . . . . . . . . . . . . . . . . . . . . . . . . . . . . . . . . . . . . . . . . . . . . . .

4  Call sign. . . . . . . . . . . . . . . . . . . . . . . . . . . . . . . . . . . . . . . . . . . . . . . . . . . . . . . . . . . . . . . . . . . . . . . . . . .

5  IMO Number . . . . . . . . . . . . . . . . . . . . . . . . . . . . . . . . . . . . . . . . . . . . . . . . . . . . . . . . . . . . . . . . . . . . . . .

6  Port of inspection . . . . . . . . . . . . . . . . . . . . . . . . . . . . . . . . . . . . . . . . . . . . . . . . . . . . . . . . . . . . . . . . . . . .

7  Date of inspection . . . . . . . . . . . . . . . . . . . . . . . . . . . . . . . . . . . . . . . . . . . . . . . . . . . . . . . . . . . . . . . . . . .

8  Action taken:

a)  Deficiencies                          b)  Action taken

. . . . . . . . . . . . . . . . . . . . . . . . . . . . . .                  . . . . . . . . . . . . . . . . . . . . . . . . . . . . . . . . . . . . . . . . . . . . .

. . . . . . . . . . . . . . . . . . . . . . . . . . . . . .                  . . . . . . . . . . . . . . . . . . . . . . . . . . . . . . . . . . . . . . . . . . . . .

. . . . . . . . . . . . . . . . . . . . . . . . . . . . . .                  . . . . . . . . . . . . . . . . . . . . . . . . . . . . . . . . . . . . . . . . . . . . .

. . . . . . . . . . . . . . . . . . . . . . . . . . . . . .                  . . . . . . . . . . . . . . . . . . . . . . . . . . . . . . . . . . . . . . . . . . . . .

. . . . . . . . . . . . . . . . . . . . . . . . . . . . . .                  . . . . . . . . . . . . . . . . . . . . . . . . . . . . . . . . . . . . . . . . . . . . .

. . . . . . . . . . . . . . . . . . . . . . . . . . . . . .                  . . . . . . . . . . . . . . . . . . . . . . . . . . . . . . . . . . . . . . . . . . . . .

. . . . . . . . . . . . . . . . . . . . . . . . . . . . . .                  . . . . . . . . . . . . . . . . . . . . . . . . . . . . . . . . . . . . . . . . . . . . .

9  Next port . . . . . . . . . . . . . . . . . .    (Date) . . . . . . . . . . . . . . . . . . . . . . . . . . . . . . . . . . . . . . . . . . . . . . .

10  Supporting documentation        ☐ No  ☐ Yes      (See attached)

Signature. . . . . . . . . . . . . . . . . . . . . . . . . . . . . . . . . . . . . . . . . . . . . . . . . . . . . . . . . .

# Appendix 16

## FORMAT FOR THE REPORT OF CONTRAVENTION OF MARPOL (ARTICLE 6)

### IMO port State control procedures (resolution A.1119(30))

(Issuing authority)                                                     Copy to:    Master

(Address)

(Telephone)

(Telefax)

(Email)

| | | |
|---|---|---|
| 1 | Reporting country | ................................................................ |
| 2 | Name of ship | ................................................................ |
| 3 | Flag of ship | ................................................................ |
| 4 | Type of ship | ................................................................ |
| 5 | Call sign | ................................................................ |
| 6 | IMO number | ................................................................ |
| 7 | Gross tonnage | ................................................................ |
| 8 | Deadweight (where appropriate) | .......................................................... |
| 9 | Year of build | ................................................................ |
| 10 | Classification society | ................................................................ |
| 11 | Date of incident | ................................................................ |
| 12 | Place of incident | ................................................................ |
| 13 | Date of investigation | ................................................................ |

14    In case of contravention of discharge provisions, a report may be completed in addition to port State report on deficiencies. This report should be in accordance with parts 2 and 3 of appendix 3 and/or parts 2 and 3 of appendix 4, as applicable, and should be supplemented by documents, such as:

    .1    a statement by the observer of the pollution;

    .2    the appropriate information listed under section 1 of part 3 of appendices 3 and 4 to the Procedures, the statement should include considerations which lead the observer to conclude that none of any other possible pollution sources is in fact the source;

    .3    statements concerning the sampling procedures both of the slick and on board. These should include location of and time when samples were taken, identity of person(s) taking the samples and receipts identifying the persons having custody and receiving transfer of the samples;

    .4    reports of analyses of samples taken of the slick and on board; the reports should include the results of the analyses, a description of the method employed, reference to or copies of scientific documentation attesting to the accuracy and validity of the method employed and names of persons performing the analyses and their experience;

    .5    if applicable, a statement by the PSCO on board together with the PSCO's rank and organization;

    .6    statements by persons being questioned;

    .7    statements by witnesses;

    .8    photographs of the slick; and

.9   copies of relevant pages of Oil/Cargo Record Books, logbooks, discharge recordings, etc.

    Name and Title *(duly authorized contravention investigation official)*

..........................................................................................................

..........................................................................................................

..........................................................................................................

..........................................................................................................

*Signature*

# Appendix 17

**COMMENTS BY FLAG STATE ON DETENTION REPORT**

Name of ship: . . . . . . . . . . . . . . . . . . . . . . . . . . . . . . . . . . . . . . . . . . . . . . . . . . . . . . . . . . . . . . . . . . . .

IMO number/call sign: . . . . . . . . . . . . . . . . . . . . . . . . . . . . . . . . . . . . . . . . . . . . . . . . . . . . . . . . . . . . . .

Flag State: . . . . . . . . . . . . . . . . . . . . . . . . . . . . . . . . . . . . . . . . . . . . . . . . . . . . . . . . . . . . . . . . . . . . . . .

Gross tonnage: . . . . . . . . . . . . . . . . . . . . . . . . . . . . . . . . . . . . . . . . . . . . . . . . . . . . . . . . . . . . . . . . . . . .

Deadweight (where appropriate): . . . . . . . . . . . . . . . . . . . . . . . . . . . . . . . . . . . . . . . . . . . . . . . . . . . . . . .

Date of report: . . . . . . . . . . . . . . . . . . . . . . . . . . . . . . . . . . . . . . . . . . . . . . . . . . . . . . . . . . . . . . . . . . . .

Report by: . . . . . . . . . . . . . . . . . . . . . . . . . . . . . . . . . . . . . . . . . . . . . . . . . . . . . . . . . . . . . . . . . . . . . . .

Classification society: . . . . . . . . . . . . . . . . . . . . . . . . . . . . . . . . . . . . . . . . . . . . . . . . . . . . . . . . . . . . . .

Recognized Organization involved: . . . . . . . . . . . . . . . . . . . . . . . . . . . . . . . . . . . . . . . . . . . . . . . . . . . . .

. . . . . . . . . . . . . . . . . . . . . . . . . . . . . . . . . . . . . . . . . . . . . . . . . . . . . . . . . . . . . . . . . . . . . . . . . . . . . . .

☐ Did you receive the notification of detention? (tick the box if the answer is 'yes')

Action taken

| a) Deficiencies | b) Cause | c) Action taken |
|---|---|---|
| . . . . . . . . . . . . . . . . . . . . . | . . . . . . . . . . . . . . . . . . . . . . . . . . | . . . . . . . . . . . . . . . . . . . . . . . . . . |
| . . . . . . . . . . . . . . . . . . . . . | . . . . . . . . . . . . . . . . . . . . . . . . . . | . . . . . . . . . . . . . . . . . . . . . . . . . . |
| . . . . . . . . . . . . . . . . . . . . . | . . . . . . . . . . . . . . . . . . . . . . . . . . | . . . . . . . . . . . . . . . . . . . . . . . . . . |
| . . . . . . . . . . . . . . . . . . . . . | . . . . . . . . . . . . . . . . . . . . . . . . . . | . . . . . . . . . . . . . . . . . . . . . . . . . . |
| . . . . . . . . . . . . . . . . . . . . . | . . . . . . . . . . . . . . . . . . . . . . . . . . | . . . . . . . . . . . . . . . . . . . . . . . . . . |
| . . . . . . . . . . . . . . . . . . . . . | . . . . . . . . . . . . . . . . . . . . . . . . . . | . . . . . . . . . . . . . . . . . . . . . . . . . . |
| . . . . . . . . . . . . . . . . . . . . . | . . . . . . . . . . . . . . . . . . . . . . . . . . | . . . . . . . . . . . . . . . . . . . . . . . . . . |
| . . . . . . . . . . . . . . . . . . . . . | . . . . . . . . . . . . . . . . . . . . . . . . . . | . . . . . . . . . . . . . . . . . . . . . . . . . . |

Additional Information: . . . . . . . . . . . . . . . . . . . . . . . . . . . . . . . . . . . . . . . . . . . . . . . . . . . . . . . . . . .

# Appendix 18

*List of instruments relevant to port State control procedures*

| Instrument | Name | IMO Body | Remark |
|---|---|---|---|
| **Resolutions** | | | |
| A.797(19) | Safety of ships carrying solid bulk cargoes | CCC | |
| A.1047(27) | Principles of minimum safe manning | MSC/HTW | |
| MEPC.104(49) | Guidelines for brief sampling of anti-fouling systems on ships | III | |
| MEPC.129(53) MEPC/Circ.472 | Guidelines for port State control under MARPOL Annex VI | III | |
| MEPC.208(62) | 2011 Guidelines for inspection of anti-fouling systems on ships | III | |
| MEPC.173(58) | Guidelines for ballast water sampling (G2) | MEPC/PPR | |
| MEPC.181(59) | 2009 Guidelines for port State control under the revised MARPOL Annex VI | MEPC/PPR | |
| MEPC.252(67) | Guidelines for port State control under the 2004 BWM Convention | MEPC/III | |
| MEPC.253(67) | Measures to be taken to facilitate entry into force of the International Convention for the Control and Management of Ships' Ballast Water and Sediments, 2004 | MEPC/PPR | |
| MEPC.259(68) | 2015 Guidelines for exhaust gas cleaning systems | MEPC/PPR | |
| MEPC.279(70) | 2016 Guidelines for approval of ballast water management systems (G8) | MEPC | Supersedes MSC.174(58) |
| MSC.159(78) | Interim guidance on control and compliance measures to enhance maritime security | MSC/III | |
| MSC.286(86) | Recommendations for material safety data sheets (MSDS) for MARPOL Annex I | PPR | |
| **Circulars** | | | |
| BWM.2/Circ.42/ Rev.1 | Guidance on ballast water sampling and analysis for trial use in accordance with the BWM Convention and Guidelines (G2) | MEPC/PPR | |
| MSC/Circ.447 | Control under SOLAS regulation I/19 – Recommendation on radar reflectors for liferafts and on training manuals | SSE | |
| MSC/Circ.592 | Carriage of dangerous goods | CCC | |
| MSC/Circ.606 | Port State concurrence with SOLAS exemptions | III | |
| MSC/Circ.635 | Tonnage measurement of certain ships relevant to the International Convention on Standards of Training, Certification and Watchkeeping for Seafarers, 1978 | SDC | |
| MSC/Circ.656 | Safety of ships carrying solid bulk cargoes | CCC | |
| MSC/Circ.811 | Identification of float-free arrangements for liferafts | SSE | |
| MSC/Circ.887 | Interpretation of the term "other strategic points" in SOLAS regulation III/50 and LSA Code section VII/7.2 | SSE | |
| MSC/Circ.890 MEPC/Circ.354 | Interim guidelines for port State control related to the ISM Code | III | |
| MSC/Circ.907 | Application of SOLAS regulation III/28.2 concerning helicopter landing areas on non ro-ro passenger ships | SSE | |

| Instrument | Name | IMO Body | Remark |
|---|---|---|---|
| MSC/Circ.918 | Guidance for port State control officers in respect of certificates of competency issued under the provision of the STCW Convention | III/HTW | |
| MSC/Circ.955 | Servicing of life-saving appliances and radiocommunication equipment under the harmonized system of survey and certification (HSSC) | III | |
| MSC/Circ.1011, MEPC/Circ.383 | Measures to improve port State control procedures | III | |
| MSC/Circ.1012 | Endorsement of certificates with the date of completion of the survey on which they are based | III | |
| MSC/Circ.1016 | Application of SOLAS regulation III/26 concerning fast rescue boats and means of rescue systems on ro-ro passenger ships | SSE | |
| MSC/Circ.1059 MEPC/Circ.401 | Procedures concerning observed ISM Code major non-conformities | III | |
| MSC/Circ.1089 | Guidance on recommended anti-fraud measures and forgery prevention measures for seafarers' certificate | III/HTW | |
| MSC/Circ.1097 | Guidance relating to the implementation of SOLAS chapter XI-2 and the ISPS Code | MSC | |
| MSC/Circ.1107 | Application of SOLAS regulation II-1/3-6 on access to and within spaces in, and forward of, the cargo area of oil tankers and bulk carriers and application of the technical provisions for means of access for inspections | CCC | |
| MSC/Circ.1113 | Guidance to port State control officers on the non-security related elements of the 2002 SOLAS amendment | MSC | |
| MSC/Circ.1117 | Guidance for checking the structure of bulk carriers | SSE | |
| MSC/Circ.1156 | Guidance on the access of public authorities, emergency response services and pilots on board ships to which SOLAS chapter XI-2 and the ISPS Code apply | MSC | |
| MSC/Circ.1586 MEPC/Circ.873 FAL.2/Circ.131 LEG.2/Circ.3 | List of certificates and documents required to be carried on board ships, 2017 | III | |
| MSC.1/Circ.1191 | Further reminder of the obligation to notify flag States when exercising control and compliance measures | MSC/III | |
| MSC.1/Circ.1196 | Means of embarkation on and disembarkation from ships | SSE | |
| MSC.1/Circ.1198 | Application of SOLAS regulation XII/6.3 on corrosion prevention of dedicated seawater ballast tanks in all types of ships and double-sided skin spaces of bulk carriers and application of the performance standard for protective coatings for dedicated seawater ballast tanks in all types of ships and double-side skin spaces of bulk carriers | SSE | |
| MSC.1/Circ.1199 | Interim guidance on compliance of ships carrying dry cargoes in bulk with requirements of SOLAS chapters II-1, III, IX, XI-1 and XII | SSE | |
| MSC.1/Circ.1208 | Promoting and verifying continued familiarization of GMDSS operators on board ships | HTW | |
| MSC.1/Circ.1221 | Validity of type approval certification for marine products | III | |
| MSC.1/Circ.1342 | Reminder in connection with shore leave and access to ships | MSC | |
| MSC.1/Circ.1235 | Guidelines on security-related training and familiarization for shipboard personnel | HTW | |
| MSC.1/Circ.1326 | Clarification of SOLAS regulation III/19 | SSE | |

| Instrument | Name | IMO Body | Remark |
|---|---|---|---|
| MSC.1/Circ.1331 | Guidelines for construction, installation, maintenance and inspection/survey of means of embarkation and disembarkation | SSE | |
| MSC.1/Circ.1402 | Safety of pilot transfer arrangements | III | |
| MSC.1/Circ.1464/ Rev.1 | Unified interpretations of the provisions of SOLAS chapters II-1 and XII, of the Technical provisions for means of access for inspections (resolution MSC.158(78)) and of the Performance standards for water level detectors on bulk carriers and single hold cargo ships other than bulk carriers (resolution MSC.188(79)), as amended by MSC.1/Circ.1507 | SSE | |
| MSC.1/Circ.1565 | Guidelines on the voluntary early implementation of amendments to the 1974 SOLAS Convention and related mandatory instruments | III | |
| MSC-MEPC.4/ Circ.1 | Retention of original records/documents on board ships | III | |
| MSC-MEPC.4/ Circ.2 | Code of good practice for port State control officers | MSC/MEPC | |
| MSC-MEPC.4/ Circ.3 | Blanking of bilge discharge piping systems in port | MSC/MEPC | |
| MSC-MEPC.5/ Circ.6 | Guidance on the timing of replacement of existing certificates by the certificates issued after the entry into force of amendments to certificates in IMO instruments | III | |
| MEPC/Circ.411 | Guidance for port State control officers on issues related to the Form of the Oil Record Book Part I | MEPC | |
| MEPC.1/Circ.479 and Corr.1 | Guidelines for port State control officers whilst checking compliance with the Condition Assessment Scheme (CAS) | MEPC/III | |
| MEPC.1/Circ.508 | Bunker delivery note and fuel oil sampling | MEPC/III | |
| MEPC.1/Circ.513 | Validity of the IOPP Certificate and Supplements issued under the current MARPOL Annex I after 1 January 2007 | MEPC | |
| MEPC.1/Circ.516 | Public access to the condition assessment scheme (CAS) database | MEPC | |
| MEPC.1/Circ.637 | Fuel oil availability and quality | MEPC | |
| MEPC.1/Circ.640 | Interim guidance on the use of the oil record book concerning voluntary declaration of quantities retained on board in oily bilge water holding tanks and heating of oil residue (sludge) | SSE | |
| MEPC.1/Circ.675/ Rev.1 | Discharge of cargo hold washing water in the Gulfs area and Mediterranean Sea area under MARPOL Annex V | MEPC | |
| MEPC.1/Circ.834 | Consolidated guidance for port reception facility providers and users | MEPC | |
| STCW.7/Circ.12 | Advice for port State control officers and recognized organizations on action to be taken in cases where not all seafarers carry certificates and endorsements meeting STCW 95 requirement after 1 February 2002 | HTW | |
| STCW.7/Circ.16 | Clarification of transitional provisions relating to the 2010 Manila Amendments to the STCW Convention and Code | MSC/HTW | |
| STCW.7/Circ.17 | Advice for port State control officers on transitional arrangements leading up to the full implementation of the requirements of the 2010 Manila Amendments to the STCW Convention and Code on 1 January 2017 | MSC/HTW | |

| Instrument | Name | IMO Body | Remark |
|---|---|---|---|
| STCW.7/Circ.24/ Rev.1 | Guidance for Parties, Administrations, port State control authorities, recognized organizations and other relevant parties on the requirements under the STCW Convention, 1978, as amended | III/HTW | |

# Appendix 19

*Port State control regimes: comparative table*
*(extracts from document III 5/5/1)*

| 1. Maritime Authorities – Members and associates | |
|---|---|
| **Paris MoU** | **27 + 1 cooperating member** |
| | Belgium, Bulgaria, Canada, Croatia, Cyprus, Denmark, Estonia, Finland, France, Germany, Greece, Iceland, Ireland, Italy, Latvia, Lithuania, Malta, Netherlands, Norway, Poland, Portugal, Romania, Russian Federation, Slovenia, Spain, Sweden and United Kingdom; Montenegro (cooperating member) |
| **Viña del Mar Agreement** | **15** |
| | Argentina, Bolivia (Plurinational State of), Brazil, Chile, Colombia, Cuba, Dominican Republic, Ecuador, Guatemala, Honduras, Mexico, Panama, Peru, Uruguay and Venezuela (Bolivarian Republic of) |
| **Tokyo MoU** | **20 members + 1 cooperating member** |
| | Australia, Canada, Chile, China, Fiji, Hong Kong (China), Indonesia, Japan, Republic of Korea, Malaysia, Marshall Islands, New Zealand, Papua New Guinea, Peru, Philippines, Russian Federation, Singapore, Thailand, Vanuatu and Viet Nam; Panama (cooperating member) |
| **Caribbean MoU** | **17 + 1 associate member** |
| | Antigua and Barbuda, Aruba (KNL), Bahamas, Barbados, Belize, Cayman Islands (UK), Curaçao (KNL),* Cuba, France, Grenada, Guyana, Jamaica, Netherlands, Saint Kitts and Nevis, Saint Lucia, Suriname, Trinidad and Tobago; Saint Vincent and the Grenadines (associate member) |
| **Mediterranean MoU** | **10** |
| | Algeria, Cyprus, Egypt, Israel, Jordan, Lebanon, Malta, Morocco, Tunisia and Turkey |
| **Indian Ocean MoU** | **21** |
| | Australia, Bangladesh, Comoros, Djibouti,* Eritrea, France, India, Iran (Islamic Republic of), Kenya, Maldives, Mauritius, Mozambique, Madagascar, Myanmar, Oman, Seychelles, South Africa, Sri Lanka, Sudan, United Republic of Tanzania and Yemen |
| **Abuja MoU** | **22** |
| | Angola, Benin, Cameroon,* Cabo Verde, Congo, Côte d'Ivoire, Democratic Republic of the Congo (DRC),* Equatorial Guinea,* Gabon, Gambia, Ghana, Guinea, Guinea Bissau, Liberia, Mauritania,* Namibia,* Nigeria, Sao Tome and Principe, Senegal, Sierra Leone, South Africa and Togo |
| **Black Sea MoU** | **6** |
| | Bulgaria, Georgia, Romania, Russian Federation, Turkey and Ukraine |
| **Riyadh MoU** | **6** |
| | Bahrain, Kuwait, Oman, Qatar, Saudi Arabia and United Arab Emirates |

---

* Pending acceptance.

| | 2. Observers |
|---|---|
| **Paris MoU** | IMO, ILO, USCG, Tokyo MoU, Caribbean MoU, Mediterranean MoU, Black Sea MoU, Riyadh MoU, Viña del Mar Agreement, Indian Ocean MoU and Abuja MoU |
| **Viña del Mar Agreement** | IMO, ILO, USCG, ROCRAM, Black Sea MoU, Caribbean MoU, Tokyo MoU and Paris MoU |
| **Tokyo MoU** | Democratic People's Republic of Korea; Macao (China); Samoa; Solomon Islands; Tonga; USCG; IMO, ILO, Black Sea MoU, Indian Ocean MoU, Paris MoU, Riyadh MoU, Caribbean MoU and Viña del Mar Agreement |
| **Caribbean MoU** | IMO, ILO, USCG, CARICOM, Paris MoU, Viña del Mar Agreement, Tokyo MoU, Anguilla (UK), Bermuda (UK), British Virgin Islands (UK), Dominica, Haiti, Montserrat (UK), Turks and Caicos Islands (UK) |
| **Mediterranean MoU** | IMO, ILO, EC, Paris MoU, Black Sea MoU and USCG |
| **Indian Ocean MoU** | IMO, ILO, Abuja MoU, Black Sea MoU, Caribbean MoU, Equasis, Ethiopia, USCG, Paris MoU, Tokyo MoU and Riyadh MoU |
| **Abuja MoU** | IMO, ILO, Mali, Burkina Faso, MOWCA, APMIAS, FAO and eight other regional PSC regimes |
| **Black Sea MoU** | IMO, ILO, Republic of Azerbaijan, USCG, Mediterranean MoU, Paris MoU, Riyadh MoU and Commission on the protection of the Black Sea against pollution |
| **Riyadh MoU** | IMO, ILO, Paris MoU, Tokyo MoU, Indian Ocean MoU, Black Sea MoU, Mediterranean MoU, Abuja MoU, Caribbean MoU, Viña del Mar Agreement, USCG, Gulf Cooperation Council (GCC) and Equasis |

| | 3. Performance review | 4. Target Inspection Rate |
|---|---|---|
| **Paris MoU** | Peer review (with Paris MoU Volunteers) and performance review (Fair Share) at every Committee meeting | The scope, frequency and priority of inspections are determined on the basis of a ship's risk profile |
| **Viña del Mar Agreement** | Performance review at every Committee meeting | 20% six-month inspection rate per country |
| **Tokyo MoU** | Performance review at every Committee meeting and introduction of peer support review | 80% annual regional inspection rate |
| **Caribbean MoU** | | 15% annual inspection rate per country within 3 years |
| **Mediterranean MoU** | | 15% annual inspection rate per country within 3 years |
| **Indian Ocean MoU** | Performance review at every Committee meeting | 10% annual inspection rate per country within 3 years |
| **Abuja MoU** | Performance review at every Committee meeting | 15% annual inspection rate per country within 3 years |
| **Black Sea MoU** | Review of the performance at every Committee meetings | 75% annual regional inspection rate |
| **Riyadh MoU** | Review of the performance at every Committee meetings | 15% annual inspection rate per country within 3 years |
| **USCG** | Review of the performance is part of the PSC Annual Report data review and is conducted annually at the Chief Inspections' Division meetings | 100% annual inspection rate per vessel, safety risk and ISPS risk matrix applied to all arriving vessels |

| 5. Relevant Instruments | | | | | | | | | | | | | | | | |
|---|---|---|---|---|---|---|---|---|---|---|---|---|---|---|---|---|
| | LL 66 | LL PROT 88 | SOLAS 74 | SOLAS PROT 78 | SOLAS PROT 88 | MARPOL 73/78/97 | STCW 78 | COLREG 72 | TONNAGE 69 | ILO 147 | ILO 147 ProT 96 | MLC 06 | AFS 2001 | CLC 69/92 | BUNKERS 2001 | BWM |
| **Paris MoU** | x | x | x | x | x | x | x | x | x | x | x | x | x | x | x | x |
| **Viña del Mar Agreement** | x | x | x | x | x | x | x | x | x | | | x | x | x | | |
| **Tokyo MoU** | x | x | x | x | x | x | x | x | x | x | | x | x | x | | x |
| **Caribbean MoU** | x | x | x | x | x | x | x | x | x | x | x | x | x | x | x | x |
| **Mediterranean MoU** | x | x | x | x | x | x | x | x | x | x | x | x | x | x | x | x |
| **Indian Ocean MoU** | x | x | x | x | x | x | x | x | x | x | x | x | x | x | x | x |
| **Abuja MoU** | x | x | x | x | x | x | x | x | x | x | x | x | x | x | | |
| **Black Sea MoU** | x | x | x | x | x | x | x | x | x | x | x | x | x | | x | x |
| **Riyadh MoU** | x | | x | x | | x | x | x | x | x | | | | | | |
| **USCG** | x | x | x | x | x | x | x | x | x | x | | | x | | | |

| | 6. Signature place and date of the Memorandum/Agreement | | 7. Official Languages |
|---|---|---|---|
| **Paris MoU** | Paris, France | 26 January 1982 | English and French |
| **Viña del Mar Agreement** | Viña Del Mar, Chile | 5 November 1992 | Portuguese and Spanish |
| **Tokyo MoU** | Tokyo, Japan | 1 December 1993 | English |
| **Caribbean MoU** | Christ Church, Barbados | 9 February 1996 | English |
| **Mediterranean MoU** | Valletta, Malta | 11 July 1997 | English |
| **Indian Ocean MoU** | Pretoria, South Africa | 5 June 1998 | English |
| **Abuja MoU** | Abuja, Nigeria | 22 October 1999 | English, French and Portuguese |
| **Black Sea MoU** | Istanbul, Turkey | 7 April 2000 | English |
| **Riyadh MoU** | Riyadh, Saudi Arabia | 30 June 2004 | Arabic and English (Official text of Memorandum is in English) |

|  | 8. Chairman 2017 | 9. Secretary and Secretariat Address | 10. Information Centre Director Address |
|---|---|---|---|
| **Paris MoU** | Mr. Brian HOGAN (Ireland) | Mr. Richard SCHIFERLI (Netherlands) Secretary General Rijnstraat 8 P.O. Box 16191, 2500 BD The Hague, the Netherlands | EMSA, Praca Europa 4, 1249-206 Lisbon, Portugal (THETIS database) |
| **Viña del Mar Agreement** | Elected for each committee | Mr. Martin Pablo RUIZ (Argentina) Av. Eduardo Madero 235, 8° piso, Of. 8:20 y 8:21 (1106) Buenos Aires, Argentina | Mr. Arnaldo Ariel Vallejos Administrator of the Information Centre of the Latin American Agreement, Prefectura Naval Argentina, Buenos Aires |
| **Tokyo MoU** | Mr. Carlos FANTA (Chile) | Mr. Hideo KUBOTA (Japan) Ascend Shimbashi 8F 6-19-19 Shimbashi Minato-ku, Tokyo 105-0004 Tokyo, Japan | Mrs. Natalia KHARCHENKO Asia-Pacific Computerized Information System (APCIS) Moscow, Russia |
| **Caribbean MoU** | Mr. Dwight GARDINER (Antigua and Barbuda) | Mrs. Jodi BARROW (Jamaica) The Office Centre Building 2nd Floor 12 Ocean Boulevard Kingston, Jamaica, W.I. | Mr. Majere AJAMBIA Database Manager Caribbean Maritime Information Centre (CMIC) Paramaribo, Suriname |
| **Mediterranean MoU** | Capt. Mark A. CHAPELLE (Malta) | Adm. Mokhtar AMMAR (Egypt) P.O.Box: 3101 746 Blue Horizon Building El Cornish St. 17th Floor Mandara, Alexandria, Egypt | Mr. Mehdi Loutfi CIMED, Acting Director Information Centre Casablanca, Immeuble Direction de la Marine Marchande, Boulevard Félix Houphouet BOIGNY, 20000, Casablanca, Morocco |
| **Indian Ocean MoU** | Ms. Beatrice Nyamoita (Kenya) | Mr. Dilip MEHROTRA (India) House No. 92, Plot No. A-8, Rangavi Estate, DABOLIM, GOA-403801, INDIA | Indian Ocean MoU Computerized Information System (IOCIS), Information Centre Pune, India |
| **Abuja MoU** | Mr. Fidele DIMOU (Congo) | Mrs. Mfon Ekong USORO (Nigeria) 1, Joseph Street, Marina Lagos, Nigeria | Mrs. Natalia KHARCHENKO Abuja MoU Information Centre (AMIS), Moscow, Russian Federation |
| **Black Sea MoU** | Mr. Hayri HASANDAYIOGLU (Turkey) | Mr. Huseyin YUCE (Turkey) Beylerbeyi Mah. Abdullahaga Cad. No:16A Kat:3 Oda:326 34676 Uskudar/Istanbul, Turkey | Mrs. Natalia KHARCHENKO Black Sea Information System (BSIS), RF PSC/FSC Directorate Burakova str., 29 Moscow, 105118, Russia |
| **Riyadh MoU** | Dr. Rashid Mohammed AL KAYUMI (Oman) | Eng Mohammed Shaban Al Zadjali (Oman) Director of the Secretariat & information Center P.O. Box 1887 Postal Code 114, Haiy Al Mina, Sultanate of Oman | Mr. Ahmed Mahmood Al Mandhari Assistant Director of the Secretariat & information Center P.O. Box 1887 Postal Code 114 Hayy Al Mina, Sultanate of Oman |
| **USCG** | ADM Karl L. Schultz Commandant, U.S. Coast Guard | Captain Matt Edwards Chief Office of Commercial Vessel Compliance (CG-CVC) U.S. Coast Guard, Stop 7501 2703 Martin Luther King Jr Ave., SE Washington, DC 20593-7501 202-372-1210 | Commander Alan Moore Chief, Foreign & Offshore Compliance Division (CG-CVC) U.S. Coast Guard, Stop 7501 2703 Martin Luther King Jr Ave., SE Washington, DC 20593-7501 202-372-1230 |

| | 11. Last and next PSCC Meetings/Conferences | 12. Website and email | 13. IMO Workshop Office bearers |
|---|---|---|---|
| **Paris MoU** | PSCC 50, Gdansk, Poland, 22-26 May 2017<br>PSCC 51, Cascais,Portugal, 7–11 May 2018<br>PSCC 52, Russian Federation, May 2019 | www.parismou.org<br>secretariat@parismou.org | Vice Chair 7th PSC Workshop at IMO |
| **Viña del Mar Agreement** | PSCC 23, Panama, 1–6 October 2017 | www.acuerdolatino.int.ar<br>ciala@prefecturanaval.gov.ar | Chairman 2nd Workshop Prefecto Mayor P. C. Escobar (Argentina) |
| **Tokyo MoU** | PSCC 28, Russian Federation September 2017<br>PSCC 28, in Hangzhou, China, 5–8 November 2019 | www.tokyo-mou.org<br>secretariat@tokyo-mou.org | |
| **Caribbean MoU** | CPCSS 22, Oranjestad, Aruba, 21–23 June 2017<br>CPSCC 23, Grand Cayman, the Cayman Islands, 27–29, 2018 | www.caribbeanmou.org<br>secretariat@caribbeanmou.org | Chairman 5th Workshop Mrs Jodi Barrow (Jamaica) |
| **Mediterranean MoU** | PSCC 19, Cyprus, 9–11 October 2017<br>PSCC 20, Georgia – September 2018 | www.medmou.org<br>secretariat@medmou.org | Chairman 3rd Workshop Admiral H. Hosni (Egypt)<br>Chairman 4th Workshop Mr. L. Vassallo (Malta) |
| **Indian Ocean MoU** | PSCC 20, Maldives 20–24 August 2017<br>PSCC 21, Australia 6–10 August 2018 | www.iomou.org<br>iomou1@dataone.in<br>iomou.sec@nic.in | Chairman 1st Workshop Capt. W.A. Dernier (South Africa) |
| **Abuja MoU** | PSCC 9, Accra Ghana, March 2018<br>PSCC 10, Benin Republic, 2019 | www.abujamou.org<br>secretariat@abujamou.org | Vice Chairman 5th IMO PSC Workshop and Chairman 6th IMO PSC Workshop Mrs. M. E. USORO (Nigeria) |
| **Black Sea MoU** | PSCC 19, Odesa, Ukraine 17-19 April 2018<br>PSCC 20, Burgas, Bulgaria third week of April 2019 | www.bsmou.org<br>bsmousecretariat@superonline.com | Vice Chairman 4th and 6th Workshop Mr. H. Yuce (Turkey) |
| **Riyadh MoU** | PSCC 15, Muscat, Oman 5–7 March 2018<br>PSCC 16, Kuwait, 28–30 Jan 2019 | www.riyadhmou.org<br>dsecretariat@riyadhmou.org | |
| **USCG** | USCG Chief, Inspections Division Meeting, April 10–12, 2018, Washington, D.C.<br><br>Next meeting TBD | https://www.dco.uscg.mil/Our-Organization/Assistant-Commandant-for-Prevention-Policy-CG-5P/Inspections-Compliance-CG-5PC-/Commercial-Vessel-Compliance/Foreign-Offshore-Compliance-Division/PSC/PortStateControl@uscg.mil | |

| 14. Concentrated inspection campaign over the last three years | | |
|---|---|---|
| | **2015** | **2016** | **2017** |
| **Paris MoU** | Crew Familiarization for Enclosed Space Entry 1 September – 30 November (Joint with the Tokyo MoU) | MLC, 2006 1 September – 30 November | Safety of Navigation, including Electronic Chart Display Information System (ECDIS) 1 September – 30 November (Joint with the Tokyo MOU) |
| **Viña del Mar Agreement** | Crew Familiarization for Enclosed Space Entry 1 September – 30 November | Cargo securing arrangement 1 September – 30 November | Safety of Navigation 1 September – 30 November |
| **Tokyo MoU** | Crew Familiarization for Enclosed Space Entry 1 September – 30 November (Joint with the Paris MoU) | Cargo securing arrangement 1 September – 30 November (coordinated with Black Sea, Indian Ocean, Mediterranean MOUs and the Viña del Mar Agreement) | Safety of Navigation, including Electronic Chart Display Information System (ECDIS) 1 September – 30 November (Joint with the Paris MoU) |
| **Caribbean MoU** | Safety of Navigation and Hours of Work or Rest 1 September – 30 November | Crew Familiarisation for Enclosed Space Entry 1 September – 30 November | Life Saving Appliances 1 September – 30 November |
| **Mediterranean MoU** | Crew Familiarization for Enclosed Space Entry 1 September – 30 November | Cargo securing arrangement 1 September – 30 November | Safety of Navigation, including Electronic Chart Display Information System (ECDIS) 1 September – 30 November |
| **Indian Ocean MoU** | Crew Familiarization for Enclosed Space Entry 1 September – 30 November | Cargo securing arrangement 1 September – 30 November | Safety of Navigation (SOLAS Ch. V) 1 September – 30 November |
| **Abuja MoU** | | | |
| **Black Sea MoU** | Crew Familiarization for Enclosed Space Entry 1 September – 30 November | Cargo securing arrangement 1 September – 30 November | Safety of Navigation, including Electronic Chart Display Information System (ECDIS) 1 September – 30 November |
| **Riyadh MoU** | Safety of Navigation 1 October – 31 December | Pilot transfer arrangement 1 September – 30 November | Crew Familiarization for Enclosed Space Entry 1 September – 30 November |

| 15. Major Training/Technical Cooperation activities | | |
|---|---|---|
| | **2015** | **2016** | **2017** |
| **Paris MoU** | Expert Training Course on the Safety Environment for PSCOs, The Hague, the Netherlands, 3 – 6 March (6 PSCOs funded by IMO)<br><br>Specialized Training on Bulk Cargoes, The Hague, the Netherlands, 15 – 18 April<br><br>Seminar 59, The Hague, the Netherlands 23 – 25 June<br><br>Seminar 60, The Hague, the Netherlands, 3 – 5 November<br><br>Expert Training the Human Element, The Hague, the Netherlands, 7 – 10 October | Expert Training Safety and Environment, The Hague, the Netherlands, 28 February – 3 March<br><br>Specialized Training on Passenger Ships, Trieste, Italy, 10 – 13 May<br><br>Seminar 61, St. Malo, France, 21 – 23 June<br><br>Expert Training the Human Element, The Hague, the Netherlands, 4 – 7 October (6 PSCOs funded by IMO)<br><br>Seminar 62, Helsinki, Finland, 8 – 10 November | Expert Training Safety and Environment, The Hague, Netherlands, 28 February – 3 March (6 PSCOs funded by IMO)<br><br>Specialised Training on Tanker Ships, The Hague, Netherlands, 21 – 24 March<br><br>Seminar 63, Cornwall, Canada, 20 – 22 June<br><br>Expert Training the Human Element, The Hague, The Netherlands, 3 – 6 October<br><br>Seminar 64, The Hague, Netherlands, 7 – 9 November |
| **Viña del Mar Agreement** | Expert mission training course on port State control jointly organized with the Tokyo MoU, Lima, Peru, 16 – 27 March (8 PSCOs funded by IMO)<br><br>Expert mission training course on port State control jointly organized with the IOMOU under AusAid, Muscat and Sohar, Oman, from 6 to 17 December<br><br>Paris MoU Eleventh Expert Training The Hague, the Netherlands, 3 to 6 March: One PSCO (Colombia)<br><br>Tokyo MoU 5th General Training Course for PSCOs: 25 August – 19 September. One PSCO (Ecuador)<br><br>Indian Ocean Expert Mission Training for PSCOs on the Human Element in Chennai, India from 30 November to 4 December 2015: One PSCO (Peru) | 6th General Training Course for PSCOs, Yokohama, Japan, 22 August – 16 September: One PSCO (Colombia)<br><br>Expert Mission Training on PSCOs, Bandar Abbas, the Islamic Republic of Iran, 5 to 16 November: One PSCO (Argentina)<br><br>Expert Training the Human Element, The Hague, the Netherlands, 4 – 7 October: One PSCO (Mexico) | Expert Mission Training Course for Port State Control Officers – Muscat and Sohar (Oman), 18 February to 02 March: One PSCO (Colombia)<br><br>Expert Training Course on Safety and Environment – The Hague, Netherlands, 28 February to 03 March: One PSCO (Chile)<br><br>7th General Training Course for PSCO – Yokohama, Japan, 14 August to 08 September: One PSCO (Brazil) |

| 15. Major Training/Technical Cooperation activities *(cont.)* | | |
|---|---|---|
| **2015** | **2016** | **2017** |
| **Tokyo MoU** | | |
| 5th General Training Course for PSCOs, Yokohama, Japan, 24 August – 18 September (7 PSCOs funded by IMO) | 6th General Training Course for PSCOs, Yokohama, Japan, 22 August – 16 September (7 PSCOs funded by IMO) | 7th General Training Course for PSCOs, Yokohama, Japan, 14 August – 8 September (6 PSCOs funded by IMO) |
| 23rd Seminar for PSCOs and the Workshop on Capacity Building for Implementation and Management of IMO Regulations, Nadi, Fiji, 13 – 17 July | 24rd Seminar for PSCOs and the Workshop on Capacity Building for Implementation and Management of IMO Regulations, Bali, Indonesia, 18 – 22 July | 25th Seminar for PSCOs and the Workshop on Capacity Building for Implementation and Management of IMO Regulations, Shanghai, China, 10 – 14 July (4 PSCOs funded by IMO) |
| Expert mission training in China (Xiamen, 23 to 26 July); Indonesia (Batam and Balikpapan, 24 August to 4 September); Viet Nam (Phu Quoc, 30 November to 4 December) and Fiji (Suva, 30 November to 4 December) | 6th Specialized Training Course for PSCOs, Tokyo, Japan, from 14 to 18 March | 7th Specialized Training Course for PSCOs (BWMC), Busan, Republic of Korea, from 13 to 16 November |
| | Expert mission training in Papua New Guinea (Port Moresby, 2 to 6 May); Malaysia (Port Klang, 30 May to 3 June); China (Dalian, 1 to 3 June); Thailand (Bangkok, 21 to 24 June); Philippines (Batangas, 14 to 25 November); Fiji (Suva, 28 November to 2 December); Peru (Callao, 28 November to 2 December) and Viet Nam (Ho Chi Minh, 5 to 9 December) | Expert mission training in Thailand (Bangkok, 25 to 27 July); Viet Nam (Hai Phong, 23 to 27 October); Peru (Callao, 13 to 17 November); the Philippines (Manila and Batangas, 13 to 24 November); Fiji (Suva, 27 November to 1 December) |
| PSCO Exchange programme (Republic of Korea to Australia, Singapore to Hong Kong (China), Australia to Canada, Hong Kong (China) to the Russian Federation, Canada to Republic of Korea, Chile to China, the Russian Federation to New Zealand and Japan to Singapore) | PSCO Exchange programme (China to the Russian Federation, Australia and the Russian Federation to Japan, Hong Kong (China) and New Zealand to Chile, and Chile and Japan to Indonesia) | PSCO Exchange programme (Australia to Peru, Chile and Fiji to Hong Kong (China), China and Peru to Australia, Hong Kong (China) and Singapore to Japan, Japan to New Zealand and the Russian Federation to Thailand) |
| Expert mission training course on port State control jointly organized with the VMA, Lima, Peru, 16 – 27 March (8 PSCOs funded by IMO) | Expert mission training course on port State control organized for the Riyadh MOU, Bahrain, 21 February – 3 March | Expert mission training course on port State control organized for the Riyadh MOU, Oman, 19 February to 2 March (5 PSCOs funded by IMO) |
| Expert mission training course on port State control jointly organized with the IOMOU under AusAid, Muscat and Sohar, Oman, from 6 to 17 December | Expert mission training course on port State control jointly organized with the IOMOU, Bandar Abbas, Iran, 5 – 16 November (7 PSCOs funded by IMO) | |

| | **15. Major Training/Technical Cooperation activities** *(cont.)* | | |
|---|---|---|---|
| | **2015** | **2016** | **2017** |
| **Caribbean MoU** | 7th Annual PSC Seminar, Cienfuegos, Cuba April 2015: Nineteen PSCOs attended<br><br>Tokyo MoU 5th General Training Course for PSCOs, 24 August 2015 to 18 September 2015: One PSCO (Jamaica)<br><br>Paris MOU Expert Training Course on Safety and Environment in The Hague, the Netherlands from March 3 – 6, 2015: One PSCO (Belize)<br><br>Expert training course jointly organized by the Viña del Mar Agreement and Tokyo MoU in cooperation with the IMO, Buenos Aires, Argentina, 16 – 27 March: One PSCO (Jamaica)<br><br>Indian Ocean Expert Mission Training Programme for PSCOs on the Human Element in Chennai, India, 30 November – 4 December: One PSCO (Grenada) | 8th Annual PSC Seminar, Fort-de-France, Martinique, April 2017: 17 PSCOs attended<br><br>Tokyo MoU 6th General Training Course for PSCOs, August to September 2016: One PSCO (Barbados)<br><br>PMOU Expert Training on Human Elements for PSCO, The Hague, Netherlands, from 4 to 7 October 2016: One PSCO (The Bahamas)<br><br>IOMOU Expert Mission Training, Bander Abbas, Iran from 5th to 16th November 2016: One PSCO (Jamaica) | 9th Annual PSC Seminar, St. John's, Antigua and Barbuda, March 13 – 15, 2017: 19 PSCOs<br><br>Riyadh MoU Expert Mission Training on Port State Control (PSC) in Muscat and Sohar, Oman, 18 February to 2 March 2017: One PSCO (St. Vincent & the Grenadines)<br><br>PMOU Expert Training on Safety and Environment for PSCOs, The Hague, Netherlands, 28 February to 3 March 2017: One PSCO (Barbados)<br><br>Tokyo MoU 7th General training course for PSCOs, Yokohama, Japan, 25 August to 17 September 2017: One PSCO (Suriname) |
| **Mediterranean MoU** | Paris Expert training course on the Safety Environment for Port State Control One PSCO (Turkey)<br><br>Expert Training Course on Port State Control jointly organized by the Vina del Mar Agreement and Tokyo MoU, Peru. One PSCO (Jordan)<br><br>The fifth General Training course for port State control officers One PSCO (Egypt)<br><br>Expert Mission Training Course on Human Element for port State control officers One PSCO (Egypt) | The Specialized Training on the Inspection of Passenger Ships in Trieste, Italy from 10 to 13 May 2016<br><br>The PSC Seminar 61 in St Malo, France from 21 to 23 June 2016<br><br>The Tokyo MoU sixth general training (GTC6), in Yokohama, Japan, from 22 August to 16 September 2016, One PSCO (Jordan)<br><br>The Expert Mission Training on Port State Control, Bandar Abbas, the Islamic Republic of Iran, 5 – 16 November 2016, One PSCO (Jordan) | Expert Mission Training on Port State Control (PSC), organized by the Riyadh MoU and in cooperation with Tokyo MoU in Oman, 19 February – 2 March<br><br>Expert Training on Safety and Environment for Port State Control Officers (PSCO), organized by the Paris MoU, The Hague, Netherlands, 28 February – 3 March<br><br>Tokyo MoU seventh General Training Course (GTC7) hosted by the Tokyo MoU in cooperation with IMO, in Japan, 14 August – 8 September |

| 15. Major Training/Technical Cooperation activities *(cont.)* | | |
|---|---|---|
| **2015** | **2016** | **2017** |
| **Indian Ocean MoU** Third Seminar for PSCOs, Kolkatta, India, 11 to 12 March (34 participants attended) | Fourth PSCO Seminar, Male Maldives, 28 August to 1 September (9 participants attended) | Fifth PSCO Seminar, Mumbai, India, 13 to 17 November (30 participants attended) |

| | 2015 | 2016 | 2017 |
|---|---|---|---|
| **Indian Ocean MoU** | Third Seminar for PSCOs, Kolkatta, India, 11 to 12 March (34 participants attended)<br><br>Expert Mission Training Course on Human Element for PSCOs, Chennai, India, 30 November – 4 December (22 participant attended, 7 PSCOs funded by IMO)<br><br>Third Expert Mission Training on PSC at Oman, under the AusAID and AMSA public Sector Linkages Programme (PSLP) in cooperation with the Tokyo MoU: 17 PSCOs from IOMOU Region attended<br><br>Tokyo MoU Expert training course on port State control jointly organized with VMA: One PSCO (Sri Lanka)<br><br>Tokyo MOU 5th General Training Course (GTC5): One PSCO (Oman funded by IMO): One PSCO (Sri Lanka funded by IOMOU) | Fourth PSCO Seminar, Male Maldives, 28 August to 1 September (9 participants attended)<br><br>Expert Mission Training on PSCOs, Bandar Abbas, the Islamic Republic of Iran, 5 to 16 November (7 PSCOs from other PSC regimes funded by IMO)<br><br>Tokyo MoU 6th General Training Course (GTC6): One PSCO (South Africa funded by IMO); One PSCO (Sudan funded by IOMOU)<br><br>Tokyo MoU 6th Specialised Training Course (STC6): One PSCO (Kenya funded by IOMOU)<br><br>Paris MoU IMO Sponsored Training Course on Human Element: One PSCO (Bangladesh funded by IMO) | Fifth PSCO Seminar, Mumbai, India, 13 to 17 November (30 participants attended)<br><br>Tokyo MoU 7th General Training Course (GTC7): One PSCO (Seychelles) funded by IMO); One PSCO (Mozambique) funded the by IOMOU<br><br>Tokyo MoU 7th Specialised Training Course (STC7): One PSCO (Sri Lanka) funded by the IOMOU<br><br>Paris MoU IMO Sponsored Training Course on Safety and Environment : One PSCO (Bangladesh) funded by the IMO<br><br>Riyadh MoU IMO Sponsored Expert Mission Training Course: One PSCO (Sudan funded by the IMO ; One PSCO (Kenya) at their own cost<br><br>Tokyo MoU 25th PSC Seminar: One PSCO (Seychelles) funded by the IOMOU |
| **Abuja MoU** | Expert Training Course for PSCOs organized by Paris MoU/IMO, 3 – 6 March 2015. The Hague, the Netherland. One PSCO (Benin)<br><br>Expert Training Course for PSCOs organized by Tokyo MoU/ Vina del Mar/IMO, 16 – 27 March. Callao, Peru: One PSCO (Cote d'Ivoire)<br><br>5th General Training Course for PSCOs organized by Tokyo MoU/IMO, 24 August – 18 September. Yokohama, Japan: One PSCO (Guinea) | 6th General Training Course (GTC6) by Tokyo MoU, 22 August to 16 September. Yokohama, Japan. One PSCO (Nigeria)<br><br>Expert Training on Human Elements for PSCOs by Paris MoU, 4 to 7 October, The Hague, Netherlands. One PSCO (Benin)<br><br>Expert mission training course on port State control (PSC), by IOMoU/Tokyo MoU/IMO, 5 to 16 November, Bandar Abbas, the Islamic Republic of Iran. One PSCO (Liberia) | Expert mission training course on port State control (PSC) organized by Riyadh MoU/ Tokyo MoU/IMO, 19 Feb to 2 March. Muscat and Sohar, Oman. One PSCO(Congo)<br><br>Expert Training on Safety and Environment for PSCOs organized by Paris MoU/IMO, 28 Feb – 3 March. The Hague, Netherlands. One PSCO(Gabon)<br><br>7th General Training Course (GTC7) organized by Tokyo MoU, 14 August to 08 September. Yokohama, Japan. One PSCO(Senegal) |

| 15. Major Training/Technical Cooperation activities *(cont.)* | | | |
|---|---|---|---|
| | **2015** | **2016** | **2017** |
| **Black Sea MoU** | Paris MOU Expert Training on the Human Element, The Hague, the Netherlands 13 – 16 October: One PSCO (Turkey)<br><br>Expert Training Course organized by Vina Del Mar Agreement & Tokyo MOU Callao, Peru 16 – 27 March: One PSCO (Turkey)<br><br>Expert Mission Training Course on Human Element for PSCOs, Chennai, India, 30 November – 4 December : One PSCO (Ukraine)<br><br>5th General Training Course for PSCOs, Yokohama, Japan, 24 August – 18 September : One PSCO (Turkey) | Paris MOU Expert Training on Safety of Environment, the Hague the Netherlands 1 – 4 March. One PSCO (Turkey)<br><br>6th General Training Course for PSCOs, Yokohama, Japan, 22 August – 16 September: One PSCO (Ukraine)<br><br>Expert Training the Human Element, The Hague, the Netherlands, 4 – 7 October. One PSCO (Georgia)<br><br>Expert Mission Training on PSCOs, Bandar Abbas, the Islamic Republic of Iran, 5 – 16 November: One PSCO (Ukraine) | Expert Mission Training Course on PSC, Muscat and Sohar, Oman, 19 February to 2 March: One PSCO (Turkey)<br><br>Paris MOU Expert Training on Safety of Environment, the Hague, the Netherlands 28 February – 03 March. One PSCO (Georgia)<br><br>7th General Training Course for PSCOs, Yokohama, Japan, 14 August – 08 September: One PSCO (Georgia)<br><br>Paris MOU Expert Training on Safety and Environment, the Hague, the Netherlands 20 – 23 March: One PSCO (Turkey) |
| **Riyadh MoU** | The 3rd Seminar for Port State Control Officers was held in Kolkata, India from 11 to 13 March: One PSCO (Saudi Arabia)<br><br>The Expert Training Course on Port State Control was held in Callao, Peru from 16 to 27 March: One PSCO (Bahrain)<br><br>The 5th General Training Course for PSCOs, Yokohama, Japan from 24 August to 18 September: One PSCO (Bahrain)<br><br>The 13th Expert Training on the Human Element was held in The Hague, the Netherlands, 13 – 16 October: One PSCO (Saudi Arabia)<br><br>The Expert Mission Training Course on Human Element for Port State Control (PSC) was held in Chennai, India from 30 November to 4 December: One PSCO (United Arab Emirates)<br><br>The 3rd Expert Mission Training Programme was held in Sultanate of Oman from 6 to 17 December: One PSCO (United Arab Emirates) and two PSCOs (Saudi Arabia) | Tokyo MoU's 6th General Training Course for Port State Control Officers (PSCOs) was held in Yokohama, Japan from 22 August to 1 September: One PSCO (Bahrain)<br><br>Paris MoU's Expert Training Course on Human Element for PSCO's was held in The Hague, the Netherlands from 4 to 7 October: One PSCO (Bahrain)<br><br>Indian Ocean's 4th Expert Mission Training Programme was held in Bandar Abbas, Iran from 5 to 16 November: One PSCO (Oman) | Expert Mission Training Course on Port State Control (PSC), organized by the Riyadh MoU in cooperation with Tokyo MoU and IMO, Muscat and Sohar, Oman, from 18 February to 2 March (6 PSCOs from Riyadh MoU and 6 PSCOs from other PSC regimes) |

| | 15. Major Training/Technical Cooperation activities *(cont.)* | | |
|---|---|---|---|
| | **2015** | **2016** | **2017** |
| **USCG** | USCG Port State Control Course, USCG Training Center Yorktown, VA, 2 week course held in February, June, and September | USCG Port State Control Course, USCG Training Center Yorktown, VA, 2 week course held in February, June September, and October | USCG Port State Control Course, USCG Training Center Yorktown, VA, 2 week course held in February, June September, and October |
| | USCG Foreign Passenger Vessel Examiner Course, USCG, USCG Cruise Ship National Center of Expertise, Miami, FL, a one week course held three times per year | USCG Foreign Passenger Vessel Examiner Course, USCG, USCG Cruise Ship National Center of Expertise, Miami, FL, a one week course held three times per year | USCG Foreign Passenger Vessel Examiner Course, USCG, USCG Cruise Ship National Center of Expertise, Miami, FL, a one week course held three times per year |
| | Liquefied Gas Carrier Inspector Course, Calhoon M.E.B.A. Engineering School, Easton, MD, a one week course held twice a year | Liquefied Gas Carrier Inspector Course, Calhoon M.E.B.A. Engineering School, Easton, MD, a one week course held twice a year | Liquefied Gas Carrier Inspector Course, Calhoon M.E.B.A. Engineering School, Easton, MD, a one week course held twice a year |
| | Chemical Tanker Safety Course, Calhoon M.E.B.A. Engineering School, Easton, MD, a one week course held twice a year | Chemical Tanker Safety Course, Calhoon M.E.B.A. Engineering School, Easton, MD, a one week course held twice a year | Chemical Tanker Safety Course, Calhoon M.E.B.A. Engineering School, Easton, MD, a one week course held twice a year |
| | Crude Oil Washing/Inert Gas System Course, Calhoon M.E.B.A. Engineering School, Easton, MD, a one week course held twice a year | Crude Oil Washing/Inert Gas System Course, Calhoon M.E.B.A. Engineering School, Easton, MD, a one week course held twice a year | Crude Oil Washing/Inert Gas System Course, Calhoon M.E.B.A. Engineering School, Easton, MD, a one week course held twice a year |
| | USCG Outer Continental Shelf Inspector Course, Shell Training facility, Robert LA, a one week course held once a year | USCG Outer Continental Shelf Inspector Course, Shell Training facility, Robert LA, a one week course held once a year | USCG Outer Continental Shelf Inspector Course, Shell Training facility, Robert LA, a one week course held once a year |

| 16. Total Number of Inspections and Detention percentages | | | | | | | | | |
|---|---|---|---|---|---|---|---|---|---|
| | **2014** | | | **2015** | | | **2016** | | |
| | Inspections | Detentions | Detention % | Inspections | Detentions | Detention % | Inspections | Detentions | Detention % |
| **Paris MoU** | 18,477 | 623 | 3.38 | 17,877 | 610 | 3.41 | 17,840 | 683 | 3.83 |
| **Viña del Mar Agreement** | 7,440 | 76 | 1.02 | 6,872 | 60 | 0.87 | 8,517 | 47 | 0.69 |
| **Tokyo MoU** | 30,405 | 1,203 | 3.96 | 31,407 | 1,153 | 3.67 | 31,678 | 1,090 | 3.44 |
| **Caribbean MoU** | 836 | 15 | 1.79 | 867 | 18 | 2.08 | 862 | 15 | 1.74 |
| **Mediterranean MoU** | 5,049 | 298 | 5.90 | 5,740 | 300 | 5.23 | 5,312 | 228 | 4.30 |
| **Indian Ocean MoU** | 6,059 | 379 | 6.26 | 6,253 | 350 | 5.60 | 6,010 | 370 | 6.16 |
| **Abuja MoU** | 2,916 | 14 | 0.48 | 2,348 | 9 | 0.38 | 1,922 | 24 | 1.25 |
| **Black Sea MoU** | 5,092 | 151 | 2.97 | 4,997 | 218 | 4.36 | 5,066 | 229 | 4.52 |
| **Riyadh MoU** | 3,859 | 44 | 1.14 | 4,165 | 32 | 0.77 | 3,381 | 26 | 0.77 |
| **USCG** | 9,232 | 143 | 1.55 | 9,265 | 202 | 2.18 | 9,390 | 98 | 0.98 |

| 17. Interregional/Global Data Exchange | | | | |
|---|---|---|---|---|
| | **IMO** | **EQUASIS** | **MoUs/USCG** | **Others** |
| **Paris MoU** | Detention data provided<br><br>Signed Data Exchange Protocol at FSI 20<br><br>Data exchange with GISIS: Live | Data provider to Equasis | Deep hyperlink to Tokyo MoU, Black Sea MoU and Abuja MoU | Agreements signed to provide data to IHS Maritime & Trade and Lloyd's List Intelligence |
| **Viña del Mar Agreement** | All inspection data<br><br>Signed data exchange agreement during the 5th IMO Workshop<br><br>Data exchange with GISIS: Under development | Data provider to Equasis | Hyperlink with Tokyo MoU<br>Paris MoU Data Interchange | Agreements signed to provide data to IHS Maritime & Trade<br><br>Agreements signed to provide data to Lloyd's List Intelligence |
| **Tokyo MoU** | All inspection data<br><br>Signed data exchange agreement during the 5th IMO Workshop<br><br>Data exchange with GISIS: Live | Data provider to Equasis | Inter-regional exchange with Black Sea MoU, Indian Ocean MoU, Paris MoU and the Viña del Mar Agreement<br><br>Inter-regional exchange with Caribbean MOU will be implemented in the near future | Data Exchange with IHS Maritime & Trade<br><br>Data Exchange with Lloyd's List |
| **Caribbean MoU** | All inspection data<br><br>Signed data exchange agreement during FSI 20<br><br>Data exchange with GISIS: Testing | Data Exchange with Equasis signed in November 2013 | Hyperlink with Paris MOU, TMOU and IOMOU | Data Exchange with IHS Maritime & Trade<br><br>Data Exchange with Lloyd's List Intelligence |
| **Mediterranean MoU** | All inspection data<br><br>Signed data exchange agreement during FSI 18 and renewed data exchange agreement during PSCWS 6<br><br>Data exchange with GISIS: Live | Data provider to Equasis | Hyperlink to Black Sea MoU and CMOU MoU | Data Exchange with IHS Maritime & Trade<br><br>Data Exchange with Lloyd's List Intelligence<br><br>Data Exchange with Genscape Vesseltracker |
| **Indian Ocean MoU** | All inspection data<br><br>Signed data exchange agreement during FSI 18 and renewed data exchange agreement during PSCWS 6<br><br>Data exchange with GISIS: Live | Data provider to Equasis<br>Hyper link to Equasis | Hyperlink to Tokyo MoU | Data Exchange with IHS Maritime & Trade<br><br>Data Exchange with Lloyd's List Intelligence |
| **Abuja MoU** | All inspection data provided<br><br>Signed Data Exchange agreement during FSI 20<br><br>Data exchange with GISIS: Live | Data Exchange with Equasis is pending | Hyperlink to Paris MoU and Indian Ocean MoU | Agreements signed to provide data to Lloyds List Intelligence<br><br>Data Exchange with IHS Maritime & Trade pending |
| **Black Sea MoU** | Detention data provided<br><br>Data exchange protocol signed during FSI 21<br><br>Data exchange with GISIS: Live | Data provider to Equasis | Hyperlink to Paris MoU, Tokyo MoU and Mediterranean MoU | Data exchange with IHS Maritime & Trade |

| 17. Interregional/Global Data Exchange *(cont.)* | | | |
|---|---|---|---|
| | **IMO** | **EQUASIS** | **MoUs/USCG** | **Others** |
| **Riyadh MoU** | All inspection data Signed data exchange agreement during the 5th IMO Workshop<br><br>Data exchange with GISIS: Under development (Final stages of testing) | Data Exchange with Equasis is under process | Working on Hyperlink with Black Sea MoU | Exploring data exchange with IHS Maritime & Trade |
| **USCG** | Detention data provided | Data provider to Equasis | | Data provider to IHS Maritime & Trade |

| | **18. Publication of Inspection Data** | **19. Publication of Detention Data** |
|---|---|---|
| **Paris MoU** | Online publication on the public website | Monthly and current detention lists on the public website |
| **Viña del Mar Agreement** | Online publication on the public website<br><br>Ship inspection search on the public website under development | Monthly detention lists on the public website |
| **Tokyo MoU** | Online publication on the public website | Online detention lists on the public website |
| **Caribbean MoU** | Online publication on the public website | Online detention lists on the public website |
| **Mediterranean MoU** | Online publication on the public website | Monthly and current detention lists on the public website |
| **Indian Ocean MoU** | Online publication on the public website | Online detention list on the public website |
| **Abuja MoU** | Online publication on the public website | Ship inspection search on the public website |
| **Black Sea MoU** | Online publication on the public website | Monthly detention list on the public website<br><br>Monthly Ship Watch List on the public website |
| **Riyadh MoU** | Online publication on the public website | Monthly and current detention lists on the public website |
| **USCG** | Online publication on the public website | Monthly detention lists on the public website |

|  | 20. Targeting System | 21. Performance System | 22. Banning/ Refusal of Access | 23. Reward System | 24. Detention Review |
|---|---|---|---|---|---|
| **Paris MoU** | Computerized Ship Targeting System | White – Grey – Black Lists | Banning/Refusal of Access | | Detention Review Panel |
| **Viña del Mar Agreement** | Computerized Ship Targeting System | | | | Detention Review (by each Member) |
| **Tokyo MoU** | Computerized Ship Targeting System | Black – Grey – White Lists | Publication of under-performing ships/Inspection at every port call | | Detention Review Panel |
| **Caribbean MoU** | Targeting System being utilized | CMOU Rating List being finalized | Refusal of Access finalized | | Detention Review Panel |
| **Mediterranean MoU** | Computerized Ship Targeting System | Under consideration | Refusal of Access | | Detention Review Panel |
| **Indian Ocean MoU** | Computerized Targeting System | Watch List | Underperforming Ship List | | Detention Review Panel |
| **Abuja MoU** | Computerized Targeting System | Under consideration | Committee approval obtained for publication of under-performing ships and detention list on website | | Detention Review Panel |
| **Black Sea MoU** | Computerized Ship Targeting System | Monthly Ship Watch List | Publication of monthly ship watch list/ subject for inspection at every port call | | Detention Review Board |
| **Riyadh MoU** | Computerized Ship Targeting System | Under consideration | | | |
| **USCG** | Computerized Ship Targeting System | Targeting Lists | Banning/Refusal of Access | Qualship 21 & Qualship E-Zero | 46 CFR 1.03 contains appeal procedures |

Notes

Notes